Optimization in Economic Theory

Optimization in Economic Theory

BY

A. K. DIXIT

OXFORD UNIVERSITY PRESS

Oxford University Press, Walton Street, Oxford OX2 6DP

OXFORD LONDON GLASGOW NEW YORK
TORONTO MELBOURNE WELLINGTON CAPE TOWN
NAIROBI DAR ES SALAAM KUALA LUMPUR
SINGAPORE JAKARTA HONG KONG TOKYO
DELHI BOMBAY CALCUTTA MADRAS KARACHI

ISBN 0 19 877103-7

First Published 1976
Reprinted 1978, 1979

Printed in Great Britain
Fletcher & Son Ltd., Norwich

Preface

The analysis of optimum choice from among a set of alternatives is a feature of most economic problems. Models of command economies naturally introduce an overall criterion function to be maximized bearing in mind the resource availability and technological possibilities. Models of market economies find it useful to formulate the behaviour of individual agents in an optimum choice framework, with each consumer maximizing utility constrained by his budget and each producer maximizing the net worth of his enterprise given its production possibilities. Between the polar cases we have mixed economies in which the government can manipulate some policy instruments but cannot achieve complete command over resource allocation. Such economies involve two stages of optimization: given each set of government policies, the other agents make their own best choices, and in considering its optimum policy choice the government has to bear in mind these responses of private agents as well as the overall resource and technology constraints.

Even in an age where most economists of recent vintage have some knowledge of mathematics, the techniques used for optimization subject to constraints are often thought to be very esoteric. This is particularly true of decisions involving time, where the choice at one point in time affects the alternatives available at later points. The problem usually arises because students meet this type of mathematics suddenly and in its full glory, and never quite recover from the experience. It is really quite feasible to begin in a much simpler way, relating the mathematics to the economics from the beginning, and progress in manageable steps to an understanding of the more advanced methods. This short book is an attempt at this programme.

Naturally, I have emphasized understanding rather than rigour. Proofs are given only when they help further understanding, and illustrative examples and applications are chosen for their economic interest and usefulness.

In each chapter, the text is followed by examples, exercises, and references for further reading. The examples are completely solved as far as the main line of reasoning is concerned, although to save space some mechanical steps of algebra or calculus are omitted. The exercises develop some of the ideas introduced in the text or the examples. As in any textbook with a mathematical content, the examples and exercises are extremely important, and are an integral part of the process of learning from the book. Suggestions for further reading include references to some basic mathematics or economics that some readers may need, and also to more advanced books and articles where interested readers can pursue the ideas that are not fully developed here. Most of the examples and applications are drawn from micro-economic theory. However, this is not meant to be a textbook in intermediate microeconomics, and is indeed probably best studied in conjunction with such a book, mathematical or otherwise.

The knowledge of economics assumed is that contained in most first-year university courses. That of mathematics is confined initially to partial derivatives and products of matrices. Quadratic forms appear in Chapter 8 and integration in Chapter 9, but in each case only the most basic properties are used. The convexity properties used in Chapters 4 and 5 are developed there as required, as are some elementary notions concerning differential equations in Chapter 11.

The text follows an internally consistent mathematical notation, but the examples and exercises often use a different notation that is common practice for the particular application being discussed. As long as the point is borne in mind, this should not cause any confusion. Similarly, I have not used boldface for vectors, to bring out the natural way in which vectors and matrices enable us to generalize results from scalar cases. The context makes the distinction clear. Finally, to save tedious repetition, I shall usually develop the general theory only for maximization problems. A simple change of sign of the criterion admits minimization problems. Where the resulting conditions are different or worth stating separately, I include exercises requiring such statements or derivations.

I am very grateful to Robert Solow for detailed comments on an earlier draft, and to Norman Ireland for reading the entire manuscript with great care. I am, of course, solely responsible for any errors that remain. Some of the examples I have used have long been part of the folklore of the subject, and many teachers rightly regard themselves as having some moral proprietary right in them. I would like to thank them, and also apologize to them.

May 1975   A. K. D.

Contents

1. Lagrange's Method

For many simple optimization problems in economics, the solution is at a point of tangency of two curves. The best known example of this is that of a consumer who chooses the amounts of two commodities on his budget line to reach the highest possible indifference curve on his indifference map. At the chosen point, the budget line is tangent to the highest attainable indifference curve. Another example is that of a producer with given resources, who can produce any combination of amounts of two goods lying on a transformation curve showing a diminishing marginal rate of transformation. Given the prices of the two goods, he would produce that combination which yields maximum revenue. In the first example the constraint curve is a straight line, while in the second the contours of equal revenue form a family of parallel straight lines. In general, both the constraint curve and the family of level curves of the objective can be non-linear. An example of this would be a planned economy with a known transformation curve, choosing a production plan to maximize a criterion of social welfare. The contours of equal social welfare would form a convex indifference map, and the production possibility schedule would be a concave curve. There will be conditions concerning permissible curvatures to be discussed later.

The general problem leads to the very familiar picture of Figure 1.1. To give an algebraic treatment, we have to define the constraint curve by an equation. Write x_1 and x_2 for the quantities of the two goods, and let the equation relating the two be written as

$$G(x_1, x_2) = c \qquad (1.1)$$

where G is a function and c a given constant. For example, in the consumer's problem the constraint has the form $p_1 x_1 + p_2 x_2 = m$ where p_1 and p_2 are the prices of the two goods and m is the money income.

Let the optimum choice be labelled $(\bar{x}_1, \bar{x}_2)$, and let the equation of the level curve of the objective function F through this point be

$$F(x_1, x_2) = v. \qquad (1.2)$$

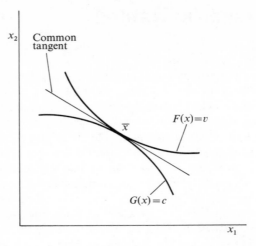

FIG. 1.1

Note that c is a datum of the problem, but that the value of v can be found only after the optimum choice is known, as $v = F(\bar{x}_1, \bar{x}_2)$.

The quantities are more compactly written as vectors arranged in vertical columns, thus

$$x = \begin{bmatrix} x_1 \\ x_2 \end{bmatrix} \qquad \text{and} \qquad \bar{x} = \begin{bmatrix} \bar{x}_1 \\ \bar{x}_2 \end{bmatrix}$$

Initially, I shall use vectors only to abbreviate lists of components. Actual operations with vectors and matrices will appear gradually.

On inspection of Figure 1.1, we have the well-known economic condition that if $\bar{x}$ is to be the optimum choice, the two curves defined by (1.1) and (1.2) should touch each other at that point. In other words, they should have the same slope there. To write this algebraically, we must find expressions for these slopes in terms of the functions F and G. Begin with the constraint curve, and consider a point $(\bar{x} + dx)$, lying on it and adjacent to $\bar{x}$, where $dx = (dx_1, dx_2)$ is an infinitesimal increment. Then dx_2/dx_1 is defined to be the slope of the curve at $\bar{x}$.

Such infinitesimal increments have the natural economic meaning of marginal changes, and their use can be given rigorous justification. But beginners sometimes make mistakes in handling infinitesimals, and it will be a useful exercise for them to rework the arguments using the standard method of calculus texts, taking finite but small changes and then going to the limit.

Since both the points being considered lie on the curve (1.1), the value of G is the same at both. In particular, the first order change dG, which is found by taking a linear approximation to G using its derivatives at $\bar{x}$, is zero. This gives

$$0 = dG = G_1(\bar{x})\,dx_1 + G_2(\bar{x})\,dx_2,$$

where G_1 and G_2 are the partial derivatives $\partial G/\partial x_j$ for $j = 1$ and 2. Of course, each is itself a function of x, and in the equation they are evaluated at $\bar{x}$. This yields

$$dx_2/dx_1 = -G_1(\bar{x})/G_2(\bar{x}),$$

the standard calculus formula for the differentiation of implicit functions. Note that if one of $G_1(\bar{x})$ and $G_2(\bar{x})$ is zero, we can still make sense of this and call it zero or infinity as the case may be. If both are zero, we are liable to have problems. Special cases may still work, but in order to be sure of the validity of general results we must confine our attention to the case where at least one of these derivatives is non-zero at $\bar{x}$.

By the same argument, the slope of (1.2) at $\bar{x}$ is $-F_1(\bar{x})/F_2(\bar{x})$. If $\bar{x}$ is the optimum choice, then the two slopes will be equal, i.e.

$$F_1(\bar{x})/F_2(\bar{x}) = G_1(\bar{x})/G_2(\bar{x}). \tag{1.3}$$

Such a condition, which necessarily holds at an optimum, is called a *necessary* condition for optimality. A condition which ensures optimality, i.e. one such that if it holds at $\bar{x}$, then $\bar{x}$ is optimum, will be a *sufficient* condition.

The left hand side, being the slope of a level curve of F, is the marginal rate of (subjective) substitution along an indifference curve of the maximand. Similarly, the right hand side is the marginal rate of transformation or technical substitution for the constraint. Thus the condition of their equality should be familiar; the two are merely expressed here in terms of the partial derivatives of the underlying functions.

Of course the point $\bar{x}$ must lie on the constraint curve, i.e.

$$G(\bar{x}_1, \bar{x}_2) = c \tag{1.4}$$

In (1.3) and (1.4) we have two equations to solve for the two unknowns, $\bar{x}_1$ and $\bar{x}_2$. The equations are usually non-linear, and we have to make careful checks before we can say whether a solution is possible or unique. Even worse, exactly the same tangency argument would have produced the same necessary conditions had we been minimizing the same function subject to the same constraint. Thus our necessary condition is far from being sufficient. However, these questions are better handled along different lines. I shall therefore neglect them for a while, and proceed assuming that $\bar{x}$ is the unique maximizer.

One important fact should be noted at this point. The number v was introduced with a warning that its value could not be known until the optimum choice had been found. Fortunately, v does not appear in (1.3) and (1.4). Thus the lack of foreknowledge does not pose any problems. We can calculate $\bar{x}$ without knowing v, and then use this to calculate the value of v.

It is useful to express (1.3) in an alternative form as

$$F_1(\bar{x})/G_1(\bar{x}) = F_2(\bar{x})/G_2(\bar{x}). \tag{1.5}$$

Write π for the common value; then we have equivalently

$$F_1(\bar{x}) - \pi G_1(\bar{x}) = 0 = F_2(\bar{x}) - \pi G_2(\bar{x}). \tag{1.6}$$

These equations can be interpreted as follows. Having defined the constant π, define a new function

$$L(x) = F(x) - \pi G(x). \tag{1.7}$$

Then (1.6) says that the partial derivatives of L are both zero when evaluated at $\bar{x}$. Now it is a well known calculus result that, if a function is maximized without any constraints, all its first order partial derivatives should equal zero at the optimum. This should be obvious from its economic meaning. For example, if a consumer is given an unlimited budget, he will choose goods until no addition to utility is possible, i.e. until the marginal utilities of all goods are reduced to zero.

Subject to a tricky point that will be taken up in Chapter 6, we see that $\bar{x}$ fulfils the necessary first order conditions for maximizing $L(x)$ without any constraints. This reduction of a constrained optimization problem to an unconstrained one is of great economic significance: the meaning will become clear in Chapter 4.

The condition (1.6) gives us an alternative method for determining $\bar{x}$. In (1.4) and (1.6), we have three equations in the three numbers $\bar{x}_1$, $\bar{x}_2$ and π. Subject to the same warnings as were given before, we can use these equations to complete the solution. As was the case with ν, we do not have to know the value of π in advance even though we began by defining it in terms of the optimum choice. In setting up the function L, we can introduce π as an 'undetermined multiplier', and obtain its value as a part of the whole process of solution.

This alternative approach is called Lagrange's method (after its inventor) for constrained optimization. The number π is called the Lagrange multiplier, and the function L is called the Lagrangean or the Lagrange expression.

This alternative approach is easy to extend to cases where there are several variables and several constraints. Clearly, two choice variables were used only to facilitate the geometric reasoning of Figure 1.1. Problems with several constraints are quite common in economics. For example, a consumer may have to budget his time as well as his income, or he may face a separate budget constraint at each point in time in drawing up his optimum consumption plan over an extended horizon. A national planner may have to ensure that his production plan does not use more of any one of several resources than the amounts available. For many of the results in the next few chapters, I shall use this last example for illustration and interpretation.

Lagrange's method is easy to extend to these problems, and the obvious generalizations turn out to be correct. Suppose there are n choice variables forming a vector x, and are subject to one constraint, $G(x) = c$, which defines a hypersurface in n-dimensional space. For the maximization of $F(x)$ we have the conditions on first-order derivatives, i.e. the *first-order conditions*

$$F_j(\bar{x}) - \pi G_j(\bar{x}) = 0 \qquad \text{for} \qquad j = 1, 2, \ldots n \qquad (1.8)$$

These n equations, together with the constraint, $G(\bar{x}) = c$, enable us to find the n components of $\bar{x}$ and the multiplier π. Next suppose there are

n choice variables and m constraints $G^i(x) = c_i$, where the functions are identified by superscripts to avoid confusion with subscripts denoting partial derivatives. We need $m < n$, for n constraints would generally reduce the feasible set to a discrete set of points, while more constraints would generally be mutually inconsistent. To extend Lagrange's method to this situation, all we have to do is to define a multiplier for each constraint. If we write π_i for the multiplier for the ith constraint, the conditions are

$$F_j(\bar{x}) - \sum_{i=1}^{m} \pi_i G_j{}^i(\bar{x}) = 0 \qquad \text{for} \qquad j = 1, 2, \ldots n \qquad (1.9)$$

where $G_j{}^i$ are the partial derivatives $\partial G^i / \partial x_j$. It is easy to verify that (1.9), and the constraining equations $G^i(\bar{x}) = c_i$, provide just the right number of equations for finding the components $\bar{x}_j$ and the multipliers π_i.

It will be convenient to express (1.9) in a more compact form using vectors. Let c be a column vector with components c_i, and G a column vector function with component functions G^i. Then all the constraints can be written together as a vector equality $G(x) = c$. Next, the partial derivatives $F_j(\bar{x})$ should be formed into a vector which I shall write as $F_x(\bar{x})$, the subscript x indicating the vector argument with respect to which the derivatives are taken. I shall make the convention that where the argument of a function is a column vector, the vector of partial derivatives will be a row vector (and vice versa; we shall meet row vector arguments later). There is a good mathematical reason for this, but the main advantage here is that it will save us from having to form frequent transposes. Similarly, for each G^i, the row vector of partial derivatives will be $G_x{}^i(\bar{x})$, and these will be stacked vertically to form an m-by-n matrix, written $G_x(\bar{x})$. The multipliers π_i will form a row vector π. Now it is easy to see, from the definition of matrix multiplication applied to (1.9), that the row vector, or 1-by-n matrix, $F_x(\bar{x})$, equals the product of the 1-by-m matrix π and the m-by-n matrix $G_x(\bar{x})$. When there was only one constraint, we had to assume that at least one $G_j(\bar{x})$ was non-zero, i.e. that $G_x(\bar{x})$ had at least one non-zero component. The generalization for more constraints is that the rows of the matrix $G_x(\bar{x})$ should be linearly independent, i.e. that it should have the maximum possible rank, m. It is easy to see that the condition for a single constraint is a special case of this: a vector on its own is linearly independent if and only if it is non-zero.

The proofs of all these generalizations are neither easy nor illuminating. Also, other more instructive methods will be used in deriving more general results in Chapters 5 and 6. I shall therefore omit the proofs here, and merely summarize the result for reference —

> If $\bar{x}$ maximizes $F(x)$ subject to the constraints $G(x) = c$, and if the matrix $G_x(\bar{x})$ has full rank, then there exists a row vector π such that

$$F_x(\bar{x}) - \pi G_x(\bar{x}) = 0 \qquad (1.10)$$

Lagrange's method provides a convenient and mechanical way to solve many economic optimization problems. We define a multiplier for each constraint, form the function L, equate its partial derivatives to zero, and solve the resulting equations and the constraints. We shall soon see ways in which this must be modified and supplemented to admit some complications that are relevant in economic problems, but the basic method will remain a valuable tool.

EXAMPLES

Example 1.1   To maximize $F(x, y) = x^\alpha y^\beta$, subject to the constraint

$$px + qy = m.$$

This will typically occur as a problem of utility maximization subject to a budget constraint, p, q being the prices of goods x, y, and m being money income. (With two variables, the (x, y) notation is simpler than (x_1, x_2).)

The first method of solution equates the slope of a level curve of the objective function to that of the constraint curve. As discussed in the text, we use the implicit function differentiation result to evaluate the former as

$$-(\partial F/\partial x)/(\partial F/\partial y) = -(\alpha x^{\alpha-1} y^\beta)/(\beta x^\alpha y^{\beta-1})$$
$$= -(\alpha/x)/(\beta/y)$$

and the latter as $-p/q$. In this example, these expressions can be found equally easily by finding an explicit equation for each curve. Thus,

$$F(x, y) = v \qquad \text{implies} \qquad y = v^{1/\beta}x^{-\alpha/\beta}$$

and along the constraint curve, we have

$$y = m/q - (p/q)x.$$

In general, explicit solutions will be much harder.

Now the optimum choice $(\bar{x}, \bar{y})$ satisfies the condition that the two slopes are equal, yielding

$$(\alpha/\bar{x})/(\beta/\bar{y}) = p/q$$

or

$$p\bar{x}/\alpha = q\bar{y}/\beta. \qquad (1.11)$$

Using the budget constraint, we can easily complete the solution

$$p\bar{x}/m = \alpha/(\alpha + \beta) \qquad \text{and} \qquad q\bar{y}/m = \beta/(\alpha + \beta)$$

The solution has the property that the budget shares are constant. This is usually not a realistic description of consumer behaviour, and better ones are available (cf. Example 1.2 and Exercise 1.4 below). However, this example has great illustrative value in many situations. Also, similar examples are somewhat more realistic in the case of production.

The second method of solution is to introduce a Lagrange multiplier π and to form the Lagrange expression

$$L(x, y) = x^{\alpha}y^{\beta} - \pi(px + qy).$$

The first-order conditions for maximization of $F(x, y)$ are found by equating each partial derivative of L to zero. Thus the optimum choice $(\bar{x}, \bar{y})$ satisfies

$$\alpha x^{\alpha-1}y^{\beta} - \pi p = 0$$
$$\beta x^{\alpha}y^{\beta-1} - \pi q = 0.$$

A useful trick for solving such equations when the constraint is linear is to multiply the first by x, the second by y, and add the two together. We have

$$(\alpha + \beta)x^{\alpha}y^{\beta} = \pi(px + qy) = \pi m.$$

If we substitute the value of π given by this in each of the equations above, we have the solution as before. Then, if we wish, we can find π in terms of the parameters of the problem. This last step is left as an exercise.

The mirror image of this problem is that of minimizing $(px + qy)$ subject to the constraint

$$x^\alpha y^\beta = z, \tag{1.12}$$

where z is a given scalar. This will typically occur as a problem of minimizing the cost of producing a target output z using factors of production x and y, when p and q are the prices of the factors, and the production function is of the product-of-powers form in (1.12), known as the Cobb-Douglas function. If there are constant returns to scale, we have $\alpha + \beta = 1$.

The method of equating the slopes shows at once that the cost-minimizing choice satisfies (1.11). However, the sum $p\bar{x} + q\bar{y}$ no longer has a known value; it is the minimum cost of production to be determined. Thus we can only say that the shares of each factor in total factor cost are constant:

$$p\bar{x}/(p\bar{x} + q\bar{y}) = \alpha/(\alpha + \beta), \; q\bar{y}/(p\bar{x} + q\bar{y}) = \beta/(\alpha + \beta)$$

With constant returns to scale, the exponents α and β are directly the factor shares. Such constancy of factor shares is sometimes an acceptable first approximation to observed producer behaviour, and this explains the popularity of the Cobb-Douglas production function.

Example 1.2   Consider another consumer choice problem with the utility function

$$F(x, y) = [\alpha x^\epsilon + \beta y^\epsilon]^{1/\epsilon}$$

The marginal rate of substitution along a level curve of this objective is

$$-\frac{(1/\epsilon)F(x, y)^{(1-\epsilon)}\alpha\epsilon x^{\epsilon-1}}{(1/\epsilon)F(x, y)^{(1-\epsilon)}\beta\epsilon y^{\epsilon-1}} = -\frac{\alpha}{\beta}\left(\frac{x}{y}\right)^{\epsilon-1}$$

If the indifference curves are to be convex to the origin, the numerical value of the above should fall as x increases or as y decreases. This needs $\epsilon < 1$. Check this by drawing some level curves for special values. For

$\epsilon = 2$ these are ellipses, which have the wrong curvature. Try -1, or $2/3$, which yields a shape that is well known in geometry. It turns out that the limiting case as ϵ goes to zero is that of Example 1.1 above.

On equating the slope above and that of the budget constraint, we have the condition for the optimum choice

$$(\alpha/\beta)\,(\bar{x}/\bar{y})^{\epsilon-1} = p/q,$$

i.e.
$$\bar{y}/\bar{x} = [(p\beta)/(q\alpha)]^{\sigma}$$

where we define $\sigma = 1/(1 - \epsilon)$. It is easy to solve this with the budget balance equation $p\bar{x} + q\bar{y} = m$ to obtain

$$p\bar{x}/m = \alpha^{\sigma}p^{-\sigma\epsilon}/(\alpha^{\sigma}p^{-\sigma\epsilon} + \beta^{\sigma}q^{-\sigma\epsilon})$$

and a similar expression for the budget share of the other good.

If we use $\epsilon = 0$, i.e. $\sigma = 1$, in this equation, we have (1.11). Thus we see that Example 1.2 is a generalization of Example 1.1. It allows the budget shares to vary systematically with the prices. For example, if ϵ is positive, the budget share of x goes to 0 as p goes to infinity, and to 1 as p goes to zero. Therefore this example has a greater potential for being a reasonable description of consumer behaviour than the one before. However, at given prices, the expenditures on each commodity are proportional to income, i.e. both income elasticities of demand are unity. This is not very reasonable, and there is still room for improvement.

EXERCISES

1.1   Solve the problem of Example 1.2 by Lagrange's method.

1.2   Generalize the two examples above to the case of n variables. (Change the notation, replacing (x, y) by $(x_1, x_2, \ldots x_n)$, (α, β) by $(\alpha_1, \alpha_2, \ldots \alpha_n)$ etc.)

1.3   It is well known in consumer theory that the exact form of the utility function is immaterial so long as the ordering of preference is preserved: F and $\hat{F}$ will serve equally well as utility functions so long as we have $F(x) \geqslant F(y)$ if and only if $\hat{F}(x) \geqslant \hat{F}(y)$. This is the case if there is an increasing function ϕ such that $\hat{F}(x) = \phi(F(x))$ for all x. To verify this, solve the consumer's problem with the utility functions

$$\hat{F}(x, y) = x^{3\alpha} y^{3\beta}$$

and $\qquad \hat{F}(x, y) = \alpha \log x + \beta \log y,$

and show that they yield the same solution as Example 1.1.

1.4   Solve the consumer's problem with the utility function

$$\alpha \log(x - x_0) + \beta \log(y - y_0)$$

where x_0 and y_0 are given numbers. Show that, provided m exceeds the value $m_0 = px_0 + qy_0$, the solution is

$$\bar{x} = x_0 + \alpha(m - m_0)/p, \bar{y} = y_0 + \beta(m - m_0)/q.$$

The parameters α and β are positive, and $\alpha + \beta = 1$. By Exercise 1.3, this involves no loss of generality. Give yourself extra credit if you can solve this by a trick without having to do any hard work.

This provides another way of generalizing Example 1.1, allowing richer possibilities for income and price elasticities. This formulation is used a great deal in practice for estimating demand systems.

FURTHER READING

Readers who need to remind themselves of the economics and the geometry of indifference curves and transformation curves can do so using any one of:

SAMUELSON, P. A. *Economics,* ninth edition, 1973, McGraw-Hill, New York, Chapter 2b and the appendix to chapter 22.

LIPSEY, R. G. *Positive Economics,* fourth edition, 1975, Weidenfeld and Nicholson, London; Chapter 4 and the appendix to chapter 15.

DORFMAN, R. *Prices and Markets,* Prentice-Hall, Englewood Cliffs, N.J., second edition, 1972, chapters 4, 5, and 7.
The last of these is a shade less elementary.

Those who need to know more about the mathematical techniques have the choice of proper mathematics books, mathematics books designed for economists, and economics books which explain mathematics along the way. In this order, I offer:

COURANT, R. and JOHN, F. *Introduction to Calculus and Analysis,* Wiley-Interscience, New York and London; Vol. I 1965, Vol. II 1974.

YAMANE, T. *Mathematics for Economists.* Prentice-Hall, New York; second edition, 1968.

ARCHIBALD, G. C. and LIPSEY, R. G. *A Mathematical Treatment of Economics,* Weidenfeld and Nicholson, London; second edition, 1973.

As far as optimization is concerned, roughly speaking, the present book begins where the book by Archibald and Lipsey ends.

I append a table of references from these books. In each case, the volume (if any) is in roman numerals, chapter in boldface, and section in arabic.

	Courant and John (I–1965, II–1974)	Yamane (1968)	Archibald and Lipsey (1973)
Derivatives	I.2.8,9; I.3.1–3	3	4–7
Partial derivatives	II.1.1–7	4	8
Chain rule	II.1.6	4.6	4.6
Taylor's theorem	II.1.7	4.5.7.1–6	
Implicit functions	II.3.1	4.7	8.6
Linear algebra	II.2.1,2	10.1–6,15	15.1–4
Matrix products	II.2.2	10.6	15.3
Quadratic forms	II.3.A1	10.12,11.7,8	–
Integration	I.2.1–3	6.1–4	13
by parts	I.3.9,11	–	Fn. p. 312

Accomplished mathematicians who wish to read a conventional proof of Lagrange's method will find one in Courant and John (II–1974), 3.7.

2. Shadow Prices

Thus far I have not given any real reason for introducing the multipliers to solve optimization problems. A problem with two variables and one constraint would be simpler without a multiplier: its use would replace the solution of two equations in two unknowns by that of three equations in three unknowns. With more variables, the multiplier makes the conditions more symmetric and easier to remember. We could have looked at cross-sections of two variables at a time and found $(n - 1)$ necessary conditions

$$F_1(\bar{x})/F_2(\bar{x}) = G_1(\bar{x})/G_2(\bar{x}), F_2(\bar{x})/F_3(\bar{x}) = G_2(\bar{x})/G_3(\bar{x}),$$
$$\ldots F_{n-1}(\bar{x})/F_n(\bar{x}) = G_{n-1}(\bar{x})/G_n(\bar{x}),$$

then proved that there were no other independent conditions, and solved these with the constraint $G(\bar{x}) = c$ for the n components of $\bar{x}$. This would be cumbersome; it would become even more so with many constraints.

But aesthetic appeal or mild simplicity are by no means the strongest reasons for using the multipliers. We would probably prefer to do without them if they did not convey some vital information about the economics of the problem. This arises in the following way.

The maximization problem has several parameters as data. The numbers c_i are obvious examples, and there will be other parameters that appear in the functions F and G^i. Economists often wish to know how the solution to the problem changes if these parameters take different values. In consumer theory, for instance, we discuss the income and substitution effects by comparing the optimum choice for different budget lines corresponding to different prices and incomes. For a producer facing given output prices, we want to know how is supply plans will change if these prices or his technology change. The general method of comparing solutions for various parameter changes is called *comparative statics,* and the importance of the Lagrange multipliers lies in the fact that they provide the answer to a very basic comparative static question.

To explain this in the simplest way, consider a problem with two choice variables and one constraint. As before, write the maximand as

$F(x)$, the constraint as $G(x) = c$, the optimum choice as $\bar{x}$, and the maximum value $v = F(\bar{x})$. Now consider a problem which differs from this only by a marginal increment dc in c, so that the constraint is $G(x) = c + dc$. Given enough regularity, we expect the solution to differ from $\bar{x}$ by a marginal amount $d\bar{x}$. The change in the maximum value is $F(\bar{x} + d\bar{x}) - F(\bar{x})$. The first order approximation to this can be calculated by successive use of first order Taylor expansions for F and G based on the derivatives at $\bar{x}$, and using the condition (1.6). We thus have, to first order,

$$dv = F_1(\bar{x})\, d\bar{x}_1 + F_2(\bar{x})\, d\bar{x}_2$$

$$= \pi\,[G_1(\bar{x})\, d\bar{x}_1 + G_2(\bar{x})\, d\bar{x}_2]$$

$$= \pi\,[G(\bar{x} + d\bar{x}) - G(\bar{x})]$$

$$= \pi\,[(c + dc) - c] = \pi\, dc$$

$$\text{or } dv/dc = \pi.$$

Thus the multiplier gives us the rate of change of the maximum attainable value of the criterion function with respect to a change in the constraining parameter. In the consumer's problem, for example, the multiplier would be the rate at which utility could be increased in response to the availability of greater money income; it would then be natural to call that multiplier the marginal utility of money income.

This generalizes very easily, and I shall discuss the general case to illustrate the ease and advantage of vectors. If $d\bar{x}$ is the column vector of increments in $\bar{x}$ corresponding to a column vector dc of increments in c, the first order change in value can be written as

$$dv = F_x(\bar{x})\, d\bar{x},$$

the matrix product of a row vector and a column vector. Then, following the steps as above and using (1.10), we have

$$dv = F_x(\bar{x})\, d\bar{x}$$

$$= \pi G_x(\bar{x})\, d\bar{x} = \pi\, dc.$$

This result is important enough to be stated separately for reference —

If dv is the first order change in the maximum value of $F(x)$ as a result of an infinitesimal increment dc in c, and π is the vector of multipliers for the constraints $G(x) = c$, then

$$dv = \pi \, dc \qquad (2.1)$$

In particular, if dc_i is the only non-zero component in dc, so that only the ith constraint changes, this reduces to $dv = \pi_i \, dc_i$. Thus π_i is the rate of change of v with respect to c_i alone; $\pi_i = \partial v/\partial c_i$.

It should be stressed that (2.1) is only the first order (linear) approximation to the change in v. For a finite change in c, we could take more derivatives and carry the Taylor expansion to higher orders to find a closer approximation. This will be done, although for a somewhat different purpose, in Chapter 8.

To illustrate and explain this result, consider a planned economy for which a production and consumption plan x is to be chosen to maximize an indicator of social welfare, $F(x)$. Suppose the various constraints $G^i(x) = c_i$ equate the different resource requirements of this plan to the availabilities of these resources. Suppose the problem has been solved and the value of the Lagrange multipliers obtained. Now suppose some power outside the economy puts an additional man-hour at its disposal. The problem can be solved afresh with the new labour constraint to determine the new pattern of production. But we know the extent of the resultant increase in social welfare without having to do this calculation: it is given by its original Lagrange multiplier, at least up to a linear approximation. We can then say that the multiplier tells us the value of the marginal product of labour for this economy in terms of its own criterion function.

Another way of looking at this is even more instructive. Suppose we use this additional man-hour for producing more output of good j alone. If dx_j is the increase in output, and the labour constraint is $G(x) = c$, then we must have $G_j(\bar{x}) \, dx_j = 1$ in order to go on satisfying the constraint when c increases by 1 (assumed to count as a small increment). Thus $dx_j = 1/G_j(\bar{x})$, and the contribution to social welfare is

$$F_j(\bar{x}) \, dx_j = F_j(\bar{x})/G_j(\bar{x}),$$

the ratio of the marginal contribution of good j to social welfare to its marginal resource requirement. At the optimum, such ratios will have been arranged to be equal for all j, since otherwise some gain in social welfare remains feasible by shifting some labour from production of a good with a lower value of this ratio to another with a higher one. Recall that it is by such verbal arguments that the proportionality of marginal utilities to the corresponding prices is established for the consumer's choice problem in elementary textbooks. Recalling (1.8), the Lagrange multiplier shows the trade-off between the constraint and the criterion. This is clearly a most important piece of economic information, and this is what establishes the importance of Lagrange's method in economics.

Now suppose this additional man-hour can only be used at some cost. The maximum cost this economy will be willing to incur in terms of its criterion is clearly equal to this multiplier, since any smaller cost will leave it with a positive net benefit from using the man-hour. In this natural sense, the multiplier represents the price that is placed on a man-hour in this economy. In the case of social welfare maximization, payments or prices expressed in terms of units of social welfare seems a strange concept. However, a minor modification brings us back on familiar ground. Consider some other resource, say land. Let the labour constraint be numbered 1 and the land constraint 2, and let π_1 and π_2 be the respective multipliers. Now suppose the economy in question is offered an additional man-hour, but asked for payment in return of the services of dc_2 units of land. The gain in social welfare from having one additional man-hour is π_1, while the loss from giving up the use of dc_2 units of land is $\pi_2 \, dc_2$. There is a net gain so long as $\pi_1 - \pi_2 \, dc_2$ is non-negative, and therefore the maximum amount of land use payment that will be offered in return for a unit of labour is (π_1/π_2). This is of course the demand price of labour for this economy, expressed relative to land. If another economy has a different trade-off on account of different resource availabilities or technology, and is willing to offer a man-hour in return for the use of a smaller amount of land, then there is the possibility of mutually advantageous trade between the two.

Of course, the internal organization of the economy need have nothing to do with prices, and the multiplier (perhaps expressed relative to another multiplier) need not equal the wage that is actually paid for each man-hour. Labour may simply be directed to various tasks in a command economy. Perhaps discriminatory pricing may be possible.

However, the multipliers remain an integral part of the outcome of the maximization problem that is solved, and they implicitly place a value on resources like labour.

However, suppose the economy does allocate resources using markets. Suppose the markets are in a state of equilibrium, where the prices are such that the demands and supplies chosen by individuals pursuing their own maximization criteria are equal in the aggregate. Now suppose an economist sets out to evaluate the performance of this economy using some given criterion. When he solves the constrained maximization problem, he will have a set of multipliers for the resource constraints. There seems little reason why the market should replicate this allocation, and the multipliers need not have any relation to the market prices of resources. But there are cases when the optimum is replicated as a market equilibrium, and the economist is tempted to say that the economy is guided to the optimum by an 'invisible hand'. This occurs in the following circumstances. Suppose the criterion has the consumers' utility levels as its only arguments, and is an increasing function of each. If the economy is competitive, with no external effects anywhere and no significant increasing returns to scale in production, and if it is possible to redistribute the initial ownership of resources as we see fit, then such an 'invisible hand' result will be true. This case has been a central concern of economic theory for a very long time. An increasing amount of attention is being paid to cases where this result cannot be true, for the conditions required are clearly very stringent. In such cases the economist must look for policies which will produce some improvement over a free market, even though the outcome may fall short of the ideal. This leads to a two-stage maximization problem in which individuals respond to policies in light of their own criteria, and the planners take these responses into account when choosing the best policy in light of theirs. In this case, we have some systematic relation, but not identity, between the planners' multipliers and the market prices, the difference being the tax or the subsidy which is the appropriate policy. Examples of such problems are (i) regulation of industries with significant increasing returns, (ii) policies concerning externalities and public goods, and (iii) tax policies which must consider a balance between equity and efficiency. I shall examine some of these in more specific contexts in later chapters.

To evoke the connection with prices, and yet maintain a conceptual distinction from market prices, the Lagrange multipliers are often called *shadow prices*.

An economic question now arises. We expect prices to be non-negative, but so far we have seen no reason why the shadow prices in our standard maximization problem should be non-negative. Clearly, relaxing a constraint should enable us to achieve a value at least as great for the criterion, but in the general statement of the problem an increase in c need not mean a relaxation of the constraint. Trivially, we could have written the constraint as $-G(x) = -c$, and an increase in the right hand side of this would mean a decrease in c. Also, not all of the constraints need be ones on resource availability. We might be maximizing the amount of investment subject to providing a given amount of consumer goods. Now an increase in this stipulated amount makes the economic constraint more severe, so a smaller amount of investment is available and the multiplier is negative. These examples show that if we want non-negative shadow prices, we must be careful to write the constraints in such a way that an increase in the right hand side does relax the restrictions on the choice being made.

There is another, more important, consideration. There may be cases in which the marginal value of a resource turns negative beyond some point. In this range, a further increase in its use will mean a lower maximum value and a negative shadow price. We have expressed the constraint as an exact equality, which forces the use of a resource even when it would have been better to leave some of it idle. If the constraint were an inequality, such as $G(x) \leq c$, we would have the freedom to do this. Of course, in adding this freedom with no other change in the problem, we assume that it is costless to leave a resource idle, which need not be the case: some resources like human brains may deteriorate faster when unused. But provided we account for such costs in the criterion, it is a good idea to allow a planner the freedom not to use some part of resources if this serves the interests of the chosen criterion. As a further argument, even the economically intuitive non-negativity of market prices would be threatened if we abandoned the assumption of costless disposability.

To admit constraints expressed by inequalities, we must develop some mathematical techniques. This will be done in Chapters 4 and 5, but one of the results is important and should be evident from the discussion above. If a part of some resource is already idle, then any increment in it will also be left idle. The maximum value of the criterion will be unchanged and the shadow price will be zero. On the other hand, a positive shadow price means that an increase in availability of the resource will increase the attainable value of the criterion. Clearly, none of the amount originally available could then have been left idle in the interests of maximization. These two arguments can be put together in the statement that, at least one of the shadow prices and the 'slack' in the use of the resource will always be zero. This general principle is one of the most important features of economic maximization problems, and it is given the name of *complementary slackness*. It will be formulated more precisely in Chapter 6, and we shall meet it again several times.

Note that writing inequality constraints in the form shown above takes care of the problem mentioned earlier, since an x which satisfies $G(x) \leqslant c$ will also satisfy $G(x) \leqslant c'$ for any c' exceeding c, ensuring that an increase in the right hand side means freer choice. A constraint which stipulates a minimum provision of some good will be of the form $G(x) \geqslant c$. In the standard form this will become $-G(x) \leqslant -c$, and an increase in the right hand side of this, i.e. a decrease in c, is again a relaxation of the constraint.

EXAMPLES

Example 2.1   Let us return to the consumer's problem of Example 1.1, and find the marginal utility of money income. The multiplier was eliminated since it was not the focus of interest there, but we do have an expression for it as part of the solution:

$$\pi = (\alpha + \beta)\bar{x}^{\alpha}\bar{y}^{\beta}/m.$$

Now we need only to substitute the values of $\bar{x}$ and $\bar{y}$ to find

$$\pi = (\alpha + \beta)\left(\frac{\alpha m}{p(\alpha + \beta)}\right)^{\alpha}\left(\frac{\beta m}{q(\alpha + \beta)}\right)^{\beta}/m$$

$$= \left(\frac{\alpha}{p}\right)^{\alpha}\left(\frac{\beta}{q}\right)^{\beta}\left(\frac{m}{\alpha + \beta}\right)^{\alpha + \beta - 1}$$

In particular, if $\alpha + \beta = 1$, the last factor equals 1, and π becomes independent of m. This makes economic sense. The case is one in which, for the scale chosen, utility shows constant returns to scale. Also, a doubling of money income at fixed prices merely leads to a doubling of both commodity quantities chosen, and therefore a doubling of utility. Therefore the marginal utility of money income is independent of the level of money income, and equal to its average utility.

Once again, it should be stressed that the whole reasoning is odd as far as consumer theory is concerned, for the particular cardinal form of the utility function does not have any special meaning and a concept such as constant returns to scale is out of place. However, welfare economics often imposes specific cardinal forms on consumers' utilities in the process of making interpersonal comparisons, and then the question becomes important. Also, for production under constant returns to scale, similar properties are true and of interest.

Example 2.2   As a step towards establishing the 'invisible hand' result mentioned in the text, consider a stage of planning where the total amounts of the various goods are known and fixed, and the only remaining question is that of distributing them among the consumers. Suppose there are I of them, labelled $i = 1, 2, \ldots I$, and that there are G goods, labelled $g = 1, 2, \ldots G$. (Recall that a different notation is being used in examples.) Let X_g be the total amount of good g, and let the amount of it allocated to individual i be x_{ig}. Each individual's utility is a function only of his own allocations,

$$u_i = U^i(x_{i1}, x_{i2}, \ldots x_{iG}) \qquad \text{for} \qquad i = 1, 2, \ldots I.$$

Social welfare is an increasing function of these utility levels

$$w = W(u_1, u_2, \ldots u_I).$$

The constraints are that for each good, its allocations to the individuals should add up to the total amount available,

$$x_{1g} + x_{2g} + \ldots + x_{Ig} = X_g \qquad \text{for} \qquad g = 1, 2, \ldots G$$

Defining Lagrange multipliers π_g for these constraints, we form the Lagrange expression

$$L = W(U^1(x_{11}, \ldots x_{1G}), \ldots U^I(x_{I1}, \ldots x_{IG}))$$

$$- \sum_g \pi_g \left[\sum_i x_{ig} \right]$$

where the range of each summation and the arguments of L are clear.

Differentiating with respect to each x_{ig} using the chain rule, we have the conditions

$$W_i U_g{}^i - \pi_g = 0, \tag{2.2}$$

where subscripts of functions indicate the appropriate partial derivatives in the usual way. They are to be evaluated at the optimum as usual, but the arguments are left out for the sake of brevity.

Now suppose the resulting numbers π_g were the prices of the respective goods in a market economy. Suppose individual i has a money income m_i, and maximizes u_i subject to the budget constraint

$$\pi_1 x_{i1} + \pi_2 x_{i2} + \ldots + \pi_G x_{iG} = m_i.$$

Defining a Lagrange multiplier λ_i for this constraint, we have the expression

$$L^i = U^i(x_{i1}, \ldots x_{iG}) - \lambda_i \sum_g \pi_g x_{ig}$$

Differentiating with respect to x_{ig}, we have the conditions

$$U_g{}^i - \lambda_i \pi_g = 0 \tag{2.3}$$

If we compare these with (2.2), we see that they coincide provided λ_i, the marginal utility of money income for individual i, equals $1/W_i$ for each i. If we have control over the ownership of resources, we can distribute it to adjust the m_i in such a way as to bring about such equalities. (It is only in exceptional cases that m_i will fail to affect λ_i, and in these cases distribution will cease to be a concern so that the problem will not arise.) Of course, this argument is of the same status as counting equations and unknowns, but like most sensible arguments of that type, it can be made rigorous. This is the 'invisible hand' result for the distribution problem.

EXERCISES

2.1   Although the choice of different cardinal forms to represent utility does not affect the optimum choice of commodities for a consumer, it does change the scale of measurement of utility and thus changes the value of the marginal utility of money income. Verify this by showing that, for the second of the functions of Exercise 1.3, we have

$$\pi = (\alpha + \beta)/m.$$

This does have the property of diminishing marginal utility of money income that acquires relevance in welfare economics.

2.2   Consider a consumer planning his consumption over two years. He will have money income m_1 during the first year and m_2 during the second. He will face prices p_1 and q_1 for goods x_1 and y_1 during the first year, and p_2 and q_2 for goods x_2 and y_2 during the second. He maximizes utility

$$u = \alpha_1 \log x_1 + \beta_1 \log y_1 + \alpha_2 \log x_2 + \beta_2 \log y_2$$

subject to two budget constraints, one for each year.

Solve this problem, and find the multipliers π_1 and π_2 for the two constraints. Examine how these depend on money income, prices, and the parameters that enter the utility function.

How much of m_2 will the consumer be willing to give up in return for being given another unit of m_1? Why would you expect institutions of borrowing and lending to develop in an economy populated by such consumers with different incomes and utility functions?

2.3   Extend the 'invisible hand' result of Example 2.2 to the following situation where the amounts of the goods to be produced are also decision variables. Suppose there are F factors of production, available in fixed amounts Z_f for $f = 1, 2, \ldots F$. If amounts z_{fg} of factor f are used in the production of good g, the outputs of the various goods are given by

$$X_g = X^g(z_{1g}, z_{2g}, \ldots z_{Fg}).$$

Maximize w as before, but now subject to constraints balancing the use and the availability of factors as well as those for the goods. From the conditions for the optimum choice, find relations between the shadow prices of goods and those factors, and interpret these relations economically.

2.4   Extend the result further to a situation where the factor supplies are also a matter for decision. Consumers supply factors, and experience disutility from doing so. Write y_{if} for the amount of factor f supplied by individual i, formulate the appropriate constraints, and proceed as in the above exercise.

FURTHER READING

The concept of the 'invisible hand' is discussed in all elementary texts, e.g. Samuelson, op. cit. (p. 11), chs. 3, 32; and Dorfman, op. cit. (p. 11), ch. 8. For a proof similar to the one here, as well as an indication of approaches that do not need derivatives, see

MALINVAUD, E. *Lectures on Microeconomic Theory*, North-Holland, Amsterdam, 1972, ch. 4.

An extremely valuable general discussion can be found in

KOOPMANS, T. C. *Three Essays on the State of Economic Science*, McGraw-Hill, New York, 1957, Essay 1.

Shadow prices are sometimes alluded to in elementary texts, e.g. Samuelson op. cit. (p. 11), pp. 775–6; and Dorfman, op. cit. (p. 11), p. 183n.

For a more detailed discussion with applications, see

HEAL, G. M. *The Theory of Economic Planning*, North-Holland, Amsterdam, 1973, Section 4.5 and Appendix A.7.

3. Maximum Value Functions

Before turning to inequality constraints, I shall discuss some other important results in comparative statics. In Chapter 2, we considered the rate of change of the maximum attainable value of the criterion with respect to the right hand side of the constraint. In other words, we recognized that the optimum choice, and therefore this maximum value, depend on the number on the right hand side of the constraint, and examined one property of this functional dependence $v(c)$, namely its derivative $v'(c)$. There is a great deal to be learnt from extending this concept further. Several other parameters enter the constrained maximization problem, and the maximum value is a function of them all. For example, the maximum utility attainable for a consumer is a function of the prices and his income; this function is called the indirect utility function. A great deal about the consumer's choice can be learnt from the properties of his indirect utility function, and sometimes it is a much better way to model his behaviour than an explicit discussion of the maximization of his ordinary (direct) utility function. This example will be taken up again in the examples and exercises at the end of this chapter. Until then, I shall consider the question in terms of the general constrained maximization problem and derive results to be applied later.

Consider first a case where the parameters affect the maximand alone. This might arise for a producer minimizing the cost of production while meeting an output target, when the parameters are the prices of the factors. Alternatively the parameters may be world prices faced by a small country wishing to maximize the value of its outputs given its resources and technology. In any case, let these parameters form a column vector b, and enlarge the list of arguments of the criterion function F to include b. Now the problem is to maximize $F(x, b)$ subject to the constraints $G(x) = c$, by choice of x. Defining a vector π of multipliers, we know that the optimum choice $\bar{x}$ satisfies

$$F_x(\bar{x}, b) - \pi G_x(\bar{x}) = 0, \qquad (3.1)$$

where now F_x is the row vector function of the partial derivatives of F with respect to the components of x, holding b constant. The maximum value is $v = F(\bar{x}, b)$, where $\bar{x}$ is itself a function of b on account of (3.1) above. Now let an increment db in b occur, and let $d\bar{x}$ and dv be the corresponding changes to first order in $\bar{x}$ and v. As in the previous chapter, we have

$$dv = F_x(\bar{x}, b) \, d\bar{x} + F_b(\bar{x}, b) \, db$$

$$= \pi G_x(\bar{x}) \, d\bar{x} + F_b(\bar{x}, b) \, db$$

But b does not enter into the constraint equations, and therefore

$$G_x(\bar{x}) \, d\bar{x} = dG = 0,$$

and

$$dv = F_b(\bar{x}, b) \, db \qquad (3.2)$$

Once again, for changes of significant size in b, we can carry the series expansion further to find closer approximations to the change in v. However, the first order result above has great interest on account of its simplicity. For it says that, in calculating the first order change in the maximum value in response to a parametric change affecting the criterion function alone, we need not worry about the simultaneous change in the choice $\bar{x}$ itself. All we have to do is to calculate the partial change with respect to the parameters, and evaluate the expression at the initial optimum choice. For the cost-minimizing producer, for example, if factor prices change the optimum factor proportions (given some substitution possibilities in production) will also change. But as far as changes in the minimum cost of production are concerned, we can (to the first order) forget about substitution and calculate as if fixed coefficients ruled.

Now let the parameters b enter the constraints as well. We can in fact subsume the right hand side vector c into the vector b, and write the general constraints in the form $G(x, b) = 0$. The previous form can then become a special case with the left hand side in the form $G(x) - c$. Now we have

$$0 = dG = G_x(\bar{x}, b) \, d\bar{x} + G_b(\bar{x}, b) \, db$$

and therefore

$$F_x(\bar{x}, b) \, d\bar{x} = \pi G_x(\bar{x}, b) \, d\bar{x} = -\pi G_b(\bar{x}, b) \, db.$$

Then, calculating as before, the change in the maximum value is

$$dv = [F_b(\bar{x}, b) - \pi G_b(\bar{x}, b)] \, db. \qquad (3.3)$$

The difference between (3.2) and (3.3) has an obvious explanation. When b affects the constraints, a change of db in it changes the value of G by an amount $G_b(\bar{x}, b)\,db$, and this acts exactly like a reduction in resource availability of an equal magnitude. The cost to the maximum value is thus $\pi G_b(\bar{x}, b)\,db$. However, once again we need not remember the induced change in $\bar{x}$.

These results suggest a more general question; namely, what happens if some of the components of x adjust to the new optimum values while the others are held fixed at their old ones? To be more precise, let the vector x be partitioned into two vectors y and z, and let the corresponding values for the optimum be $\bar{y}$ and $\bar{z}$. We want to compare the response of v to a change in b when we allow the whole of $\bar{x}$ to adjust optimally, with that when we hold $\bar{y}$ fixed and allow only $\bar{z}$ to adjust. Of course, enough components must be flexible to ensure that the constraints can be met; this needs at least as many flexible components as there are constraints.

Let us first examine the question using Lagrange's method. In the first case, when all components are free, we can rewrite the conditions (3.1) in the partitioned notation as

$$F_y(\bar{y}, \bar{z}, b) - \pi G_y(\bar{y}, \bar{z}, b) = 0$$
$$F_z(\bar{y}, \bar{z}, b) - \pi G_z(\bar{y}, \bar{z}, b) = 0 \tag{3.4}$$

and (3.2) becomes

$$dv = [F_b(\bar{y}, \bar{z}, b) - \pi G_b(\bar{y}, \bar{z}, b)]\,db. \tag{3.5}$$

When only z adjusts, we are solving the problem of maximizing $F(\bar{y}, z, b)$ subject to the conditions $G(\bar{y}, z, b) = 0$. Defining a vector of multipliers λ, we have the conditions

$$F_z(\bar{y}, \bar{z}, b) - \lambda G_z(\bar{y}, \bar{z}, b) = 0 \tag{3.6}$$

and then

$$dv = [F_b(\bar{y}, \bar{z}, b) - \lambda G_b(\bar{y}, \bar{z}, b)]\,db. \tag{3.7}$$

It is not clear whether (3.7) and (3.5) will give the same answer. A sufficient condition for this will be $\pi = \lambda$, which is cumbersome to prove, and subject to pedantic-looking qualifications.

I have set up this inconclusive argument to demonstrate by contrast the power of an alternative approach. This method relies directly on the definition of an optimum, rather than on necessary conditions in terms of derivatives. Quite simply, we use the inequality that the value of the criterion corresponding to the optimum choice must be at least as great as that corresponding to any feasible choice. In the present case, this approach tells us very easily when the desired result is valid, and also gives us the meaning of the qualifications that arise.

To formulate this, it is useful to give explicit recognition to the fact that the optimum value and the corresponding choices are all functions of the parameters. Thus, when both y and z are free, let $v = V(b)$ be the maximum value, and $\bar{y} = Y(b)$ and $\bar{z} = Z(b)$ the corresponding choices, thus

$$V(b) = F(Y(b), Z(b), b) \tag{3.8}$$

When y is fixed, we must admit it as another vector of parameters. For the problem when only z is free, write $V(y, b)$ as the maximum value, and $Z(y, b)$ as the optimum choice. Then

$$V(y, b) = F(y, Z(y, b), b) \tag{3.9}$$

The use of the same function symbols in the two cases should not cause any confusion, since in each case the arguments will be written explicitly to make it clear which case is intended.

Now the point on which the argument hinges is quite simply that the choice $(y, Z(y, b))$ satisfies the constraints which would apply if both sets of variables were free, namely $G(y, z, b) = 0$. As $V(b)$ is the optimum value for that problem, we must have

$$V(y, b) \leqslant V(b) \tag{3.10}$$

For one particular case, namely when y is held fixed at just the value it would have taken had it been free, the two solutions coincide, and therefore

$$V(Y(b), b) = V(b) \tag{3.11}$$

Now consider an increment db in b. Although $\bar{y} = Y(b)$ was the optimum choice for that set in the original problem, it need not remain so when the parameters change to $(b + db)$. We therefore have

$$\left.\begin{array}{l} V(\bar{y}, b) = V(b) \\ V(\bar{y}, b + db) \leqslant V(b + db) \end{array}\right\} \tag{3.12}$$

and subtracting,

$$V(\bar{y}, b + \mathrm{d}b) - V(\bar{y}, b) \leqslant V(b + \mathrm{d}b) - V(b). \qquad (3.13)$$

This is the general result for the comparison of changes in value when some of the choice variables are held fixed and those when they are all free. To see the earlier question in this context, consider the case when there is only one parameter b. This is illustrated in Figure 3.1. As (3.12) requires, the curve showing the maximum value as a function of b when y is held fixed at $\bar{y}$ lies everywhere below the corresponding curve for the case when y is free, and the two coincide at the particular value of b

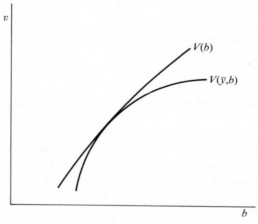

FIG. 3.1

for which $\bar{y}$ happens to be the optimum choice. If both functions are differentiable, they must be mutually tangential at this point, as in Figure 3.1. Then

$$V_b(\bar{y}, b) = V_b(b). \qquad (3.14)$$

where V_b is simply another notation for the derivative V'.

To prove this algebraically, regard $\mathrm{d}b$ as a finite change, and divide both sides of (3.13) by it. The direction of the inequality is maintained if $\mathrm{d}b$ is positive, and reversed if it is negative. In each case, proceed to the limit as $\mathrm{d}b$ goes to zero. This gives two weak inequalities pointing in opposite directions between the two sides of (3.14), thus establishing the equation provided the derivatives exist. If b is a vector, this can be done separately for each of its components, establishing equations like (3.14) for each corresponding pair of partial derivatives of the two functions, i.e. the vector equality for the derivatives as whole vectors. This is the kind of result that we have been trying to find; it equates the rates of change of the maximum value with respect to the parameters, irrespective of whether all variables are free or some of them are fixed.

However, we shall see later that maximum value functions may fail to have derivatives, i.e. they may have different slopes at a point depending on whether we approach it from the left or the right. This can happen even when all the underlying criterion and constraint functions are differentiable. This gives rise to a possibility like that shown in Figure 3.2, where (3.12) are satisfied, but an equality

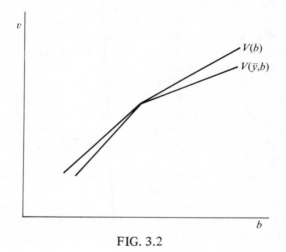

FIG. 3.2

like (3.14) cannot be inferred from them. The best we can do is to define separate 'left-handed' and 'right-handed' derivatives depending on whether db approaches zero from positive or negative values, and establish inequalities which say that the upper curve is steeper to the right and flatter to the left, i.e. in obvious notation

$$V_b(b)_+ \geqslant V_b(\bar{y}, b)_+, \quad V_b(b)_- \leqslant V_b(\bar{y}, b)_-. \tag{3.15}$$

It is in just such a case that equations like (3.5) and (3.7) will be valid but will have different values for the Lagrange multipliers.

In discussing applications, I shall assume differentiability of the maximum value function where appropriate without rigorous justification. One such application has become famous in economics, since it was the first instance to be discovered of the general properties discussed in this chapter. Cost curves are found by minimizing the cost of producing a stipulated amount of output, and drawing the value of the minimized cost as a function of the output. Marginal cost is then the slope of this curve as the output parameter changes. In the long run, all factors are at the firm's choice, while in the short run, only a subset of them can be chosen freely. For each short run cost curve, there will be a level of output at which the fixed factors happen to be at their long run optimum levels. Then (3.14) tells us that at this output level, the short run marginal cost will equal the long run marginal cost, and then the corresponding average cost curves will touch each other. Further, from (3.12), we see that the long run total (or average) cost curves must lie everywhere below the corresponding short run ones (remember that this is a minimization problem). If we repeat this for all short run cost curves, the long run cost curve will emerge as the envelope of the family of short run cost curves corresponding to different levels of the fixed factors. Hence, in fact, the general result is often called the 'envelope theorem'. It has also been called the Wong-Viner theorem, after the two who stumbled upon it while studying properties of cost curves.

The first order nature of this result should be stressed. Higher order derivatives of the two functions $V(b)$ and $V(\bar{y}, b)$ will not in general be equal. We see from Figure 3.1 that the latter must have a greater downward curvature, i.e. it must be more concave. This has some important implications, and I shall have occasion to refer to them in Chapter 8.

The next natural comparative static question concerns the response of $\bar{x}$ itself to changes in the parameters. Results like the negativity of the own substitution effect belong to this category. As $\bar{x}$ is found from (1.10) and the constraints, we have to find how these change when the parameters change. This can be done by differentiation, but (1.10) itself has first order derivatives, and changes in it will bring in second order ones. I shall postpone discussion of changes in $\bar{x}$ for two reasons.

First, the readers will in the meantime have had more practice of mathematical reasoning and will be able to understand the arguments and the results more easily. Secondly, the techniques to be developed in the next two chapters will often provide ways of answering comparative static questions more easily without resorting to second order derivatives in many important cases. The direct reasoning used in this chapter has already provided an example of the power and simplicity of such methods, and there is much to be said for developing them further and using them more frequently.

E X A M P L E S

Example 3.1   As an illustration of the envelope theorem on its home ground, consider the following situation. In the short run, if a plant designed for capacity k is to be used for producing output q, the cost is given by

$$C(q, k) = a + \tfrac{1}{2}bk^2 [1 + (q/k)^4].$$

In the long run, k is variable. In order to produce output q in the long run at the minimum cost, the producer will choose k so as to minimize $C(q, k)$. The first order condition for this is

$$C_k(q, k) = b[k - q^4 k^{-3}] = 0$$

and the second order derivative

$$C_{kk}(q, k) = b[1 + 3(q/k)^4]$$

is positive, thus ensuring a minimum. This yields $k = q$, i.e. capacity should be chosen equal to the long run output level planned. Then the long run cost curve is found by substituting this value as

$$C(q) = a + bq^2.$$

Now the short-run marginal cost is

$$C_q(q, k) = \tfrac{1}{2}bk^2 \cdot 4q^3/k^4 = 2bq^3/k^2,$$

and the long-run marginal cost is

$$C'(q) = 2bq.$$

The two are clearly equal when $k = q$.

It is a useful exercise to plot these functions, and the associated average cost curves, to scale on graph paper. To simplify paper-and-pencil calculations, take $a = 240$, $b = 15$, and try values $k = 2, 3, 4, 5$ and 6. If using a slide rule or a calculator, experiment with your own numbers.

Example 3.2   The most important development in this chapter has been the introduction of the idea of regarding the maximum value of the criterion as a function of the parameters of the problem. Such functions convey a lot of economically useful information about the optimization problem under study, and have several important applications. This example is designed to illustrate some such applications in consumer theory.

For a consumer maximizing utility subject to the budget constraint $px = m$, where p is a row vector of prices and m is money income, the maximum utility he can achieve is a function of p and m. This is called the *indirect utility function*. Write it as $V(p, m)$. Some properties of it are evident; for example, changing all prices and income in the same proportion leaves the feasible commodity bundles x unchanged, and thus does not affect the maximum attainable utility level; thus $V(p, m)$ is homogeneous of degree zero in its arguments. Some other properties will be studied later. The feature of particular interest at this point is the application of the comparative static results derived so far. Write V_m for the partial derivative $\partial V/\partial m$, and V_j for $\partial V/\partial p_j$. If, following the standard practice in consumer theory, we denote the Lagrange multiplier for the problem by λ, we know from the shadow price interpretation that

$$\lambda = V_m(p, m). \tag{3.16}$$

We can also apply (3.3). If all prices except the jth are fixed, we can find the rate of change of V with respect to p_j, i.e. V_j, from (3.3) as

$$V_j(p, m) = -\lambda \partial(px)/\partial p_j = -\lambda x_j,$$

evaluated at the optimum choice. Of course the utility-maximizing choice defines the demand functions, $x_j = D^j(p, m)$. Thus we have

$$D^j(p, m) = -V_j(p, m)/V_m(p, m). \tag{3.17}$$

This is a useful and important result. If we are given the consumer's utility function and asked to find the resultant demand functions, we have to carry out the whole constrained optimization solution, which is a messy task even in the simplest cases. On the other hand, if we are given his indirect utility function, we can find the demand functions by differentiation alone. Thus it is much simpler to summarize our information about consumers by means of indirect utility functions. Particularly in models in which the consumers are only one part of the story, the consequent economy of effort and of notation makes a great deal of difference. Some such applications will be developed in the subsequent chapters.

Next consider the mirror image problem mentioned before, where the consumer is seen as minimizing the expenditure necessary for attaining a given target utility level. The minimum value that results is now a function of the prices and of the utility level. This is called the *expenditure function*, written $E(p, u)$. Keeping u fixed and changing all prices in the same proportion will change the necessary expenditure by that proportion, and therefore the expenditure function is homogeneous of degree one in p for every fixed u. Once again, other properties will be developed later; once again, first order changes in its value tell us about demand functions.

In notation analogous to that used above, if μ is the Lagrange multiplier, we have

$$\mu = E_u(p, u) \tag{3.18}$$

In this case, price changes do not affect the constraint, and we can use (3.2). This gives

$$E_j(p, u) = x_j$$

evaluated at the optimum. Cost-minimizing commodity choices for a given utility level are the *compensated* demand functions, $C^j(p, u)$. The process is as if, following any price change, the consumer is compensated by changing his money income just enough to leave him on the same indifference curve. This is done in the two-good case by sliding the budget tangentially to the indifference curve, in order to isolate the substitution effect of a price change. Now we have shown

$$C^j(p, u) = E_j(p, u). \tag{3.19}$$

This expression is even simpler than that for the (uncompensated) demand functions above, and is often more useful. Its applications will be taken up later.

Example 3.3   Since the vector c of the right hand sides of the constraints can be subsumed in the vector of parameters b used in this chapter, it should be possible to derive (2.1) as a special case of (3.3). To do this, let us identify b and c, and consider the special case where $G(x, b) = G(x) - c$. Now the partial derivative of the ith component function with respect to c_j is -1 if $i = j$ and zero otherwise; thus the matrix G_b becomes $-I$ where I is the (m-by-m) identity matrix. The maximand does not involve c. Therefore (3.3) becomes

$$dv = [0 - \pi(-I)] \ dc = \pi \ dc,$$

which is (2.1). It is common to write the constraints $G(x) = c$ in the form $G(x) - c = 0$. Then the Lagrange expression (1.7) can be written as

$$L(x) = F(x) - \pi[G(x) - c]. \tag{3.20}$$

This is often useful in theoretical developments I shall not discuss, but the practical benefit is that (2.1) and (3.3) can be stated in a simple form: the first order derivatives of the maximum value with respect to the parameters are equal to the corresponding partial derivatives of the Lagrange expression, evaluated at the optimum.

EXERCISES

3.1   Give details of the limit arguments used in deriving (3.14) and (3.15).

3.2   Consider a producer who uses a vector of inputs x to produce a given amount of output y according to a production function $y = F(x)$. He faces prices w for these inputs. Define his minimum cost of production as a function of w and y, called the *cost function*, $C(w, y)$. Derive his factor demands for achieving minimum-cost production in terms of the derivatives of the cost function. Interpret the Lagrange multiplier for the minimization problem.

    Now suppose he faces a price p for output, and chooses its quantity to maximize profit. What further conditions emerge? If the *profit function* is defined as the maximum value of profit regarded as a function of all prices, how can the producer's supply curve for output be derived from it?

3.3 For the second case in Exercise 1.3, show that the indirect utility function is

$$V(p, q, m) = \alpha \log \alpha + \beta \log \beta - (\alpha + \beta) \log (\alpha + \beta)$$
$$+ (\alpha + \beta) \log m - \alpha \log p - \beta \log q,$$

and that for the case of Example 1.2, it is

$$V(p, q, m) = m(\alpha^\sigma p^{-\sigma\epsilon} + \beta^\sigma q^{-\sigma\epsilon})^{1/(\sigma\epsilon)}.$$

In each case, find the corresponding expenditure function. Generalize these expressions to the case of n choice variables, with proper notational changes.

3.4 For the production function

$$y = Ax_1^{\alpha_1} x_2^{\alpha_2} \ldots x_n^{\alpha_n},$$

show that the cost function is

$$C(w, y) = \gamma(y/A)^{1/\gamma} (w_1/\alpha_1)^{\alpha_1/\gamma} \ldots (w_n/\alpha_n)^{\alpha_n/\gamma}$$

where $\gamma = \sum_{j=1}^n \alpha_j$. If $\gamma < 1$, calculate the corresponding profit function. What will go wrong if $\gamma = 1$, i.e. if there are constant returns to scale in production?

FURTHER READING

For more on cost curves and their envelope properties, see Samuelson op. cit. (p. 11), ch. 24; Lipsey, op. cit. (p. 11), ch. 18; and Dorfman, op. cit. (p. 11), ch. 3. For the story of the discovery of the Wong-Viner theorem, see

VINER, J. 'Cost Curves and Supply Curves', reprinted in *Readings in Price Theory*, (eds. G. J. Stigler and K. E. Boulding), Irwin, Homewood, Ill., 1952.

Unfortunately, no textbook treats the indirect utility, expenditure, cost and profit functions at all systematically. I hope that the treatment here and in later chapters goes a small way towards filling this large gap. Some references dealing with particular applications will appear later. A general and definitive analysis of production theory with applications, long awaited in published form, is

McFADDEN, D. L. 'Cost, Revenue and Profit Functions', University of California, Berkeley, Working Paper, 1970.

For an extensive survey of applications, with the basic theory, see

DIEWERT, W. E. 'Applications of duality theory', in *Frontiers of Quantitative Economics*, Vol. II, eds. M. D. Intriligator and D. A. Kendrick, North Holland, Amsterdam, 1974, pp. 106–71.

4. Inequality Constraints

The discussion of shadow prices in Chapter 2 pointed to a major defect of theories of optimization which use constraints in equation form: they force the use of resources even when it is undesirable to do so. The methods developed in this chapter and the following two chapters remove this flaw, and thus add a lot of economic relevance to the theory. They do so in another way, too. In previous chapters, all functions were supposed to have derivatives with respect to all arguments. It is often claimed that functions appropriate to economic analysis are unlikely to be smooth enough. The results to come are valid for continuous functions, and therefore more general. This lets us remove differentiability from the list of assumptions essential to the theory, and put it in its proper role of a convenient approximation, to be used only when it does no great harm to the reality.

Finally, the mathematics we need here is simple analytic geometry — in particular the equations of straight lines and planes. This is an important advantage, for it is undeniably simpler to multiply and add numbers and compare magnitudes than it is to differentiate.

Let us begin with two variables and one constraint. The familiar picture of Figure 1.1 is easily modified to allow for inequalities, and leads to Figure 4.1. In the usual case, both F and G are increasing functions of x, so the choice of x can be on the constraint curve or below it, as in the shaded area $\mathscr{A}$. It is convenient to contrast this with the set of points yielding unattainable, or at best just attainable, values of the criterion; this is the shaded area $\mathscr{B}$ on and above the level curve through the optimum choice $\bar{x}$. If the two curves have straight line segments, the optimum choice may not be unique and $\mathscr{A}$ and $\mathscr{B}$ may have points in common besides $\bar{x}$; I shall return to such problems later. The main point to be noted here is that any points common to the two areas must be on their boundaries. There can be no points in $\mathscr{A}$ with $F(x) > v$, nor can there be points in $\mathscr{B}$ with $G(x) < c$; for in either case we would be able to find an improvement on $\bar{x}$, contrary to our definition of it as the optimum choice.

I have used the term 'boundary' in an intuitive geometric way, and this will suffice for much of our work. But it can be misleading, and a

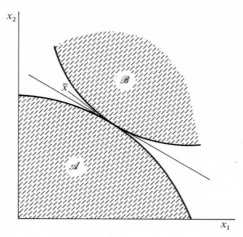

FIG. 4.1

better definition will help. A point is *interior* to a set if it is surrounded for some distance by points of that set. Thus a point s in a set $\mathscr{S}$ will be an interior point if there is a positive number r such that all points of the space within distance r of s are also in $\mathscr{S}$. In the plane, such points will form a disc of radius r centred on s. Then, a point which is interior neither to $\mathscr{S}$ nor to the rest of the space will be called a *boundary* point of $\mathscr{S}$. Thus s will be a boundary point of $\mathscr{S}$ if, for any positive r, we can find points of $\mathscr{S}$ as well as points not in $\mathscr{S}$ within distance r of s. Any point x for which $F(x) > F(\bar{x}) = v$ will be an interior point of $\mathscr{B}$ so long as F is continuous, and any x for which $G(x) < c$ will be an interior point of $\mathscr{A}$ so long as G is continuous. This minimum assumption of regularity will be maintained in all that follows. Further, so long as F and G are continuous and the set $\mathscr{A}$ is bounded, it can be proved rigorously that the problem of maximizing $F(x)$ subject to $G(x) \leqslant c$ has a solution. Existence will not present a problem except possibly in Chapters 10 and 11.

The usual assumption of a diminishing marginal rate of transformation corresponds to the requirement that the set $\mathscr{A}$ should be *convex*, i.e. given any two points in it, the whole of the straight line segment joining them should also lie within the set. Let the vector co-ordinates of the two points be x and x'. Then, as those of their midpoint are $(x + x')/2$, those of the entire straight line segment joining them can be traced out by $(\delta x + (1 - \delta)x')$, with the number δ ranging over the interval $0 \leqslant \delta \leqslant 1$. This enables us to define a convex set in terms of analytic geometry, and will be used frequently.

Similarly, the assumption of a diminishing marginal rate of substitution corresponds to the set $\mathscr{B}$ being convex. If both assumptions are made, the consequence, as in Figure 4.1, is that the sets lie one on each side of their common tangent at $\bar{x}$. Suppose the equation of this tangent is

$$\theta_1 x_1 + \theta_2 x_2 = d \tag{4.1}$$

For this to be a meaningful equation, θ_1 and θ_2 cannot both be zero, and for the line to pass through $\bar{x}$, we must have

$$\theta_1 \bar{x}_1 + \theta_2 \bar{x}_2 = d \tag{4.2}$$

For all points x on one side of the line, the value of the expression in (4.1) will exceed d, and for all those on the other side, it will fall short of d. Since the line is not altered if we multiply both sides of its equation by the same non-zero number, we can choose the sign of this number to ensure that

$$\theta_1 x_1 + \theta_2 x_2 \begin{cases} \leqslant d & \text{for all } x \text{ in } \mathscr{A} \\ \geqslant d & \text{for all } x \text{ in } \mathscr{B} \end{cases} \tag{4.3}$$

As Figure 4.1 is drawn, θ_1, θ_2 and d will all be positive when this is done. The economic reason for this will soon become clear.

The results generalize very easily. In a space of any dimension, given two convex sets which have only boundary points in common, we can find a hyperplane such that the sets lie one on each side of it, or in other words, the hyperplane *separates* the sets. A hyperplane has a linear equation, $\theta x = d$, where θ is a non-zero row vector. Then for all points in one of the sets, we will have $\theta x \leqslant d$, and for the other, $\theta x \geqslant d$. This is quite obvious from geometric intuition, and I shall leave it to the reader to convince himself by drawing a few pictures, and omit the proof. However, there is a small complication to be resolved. A straight line segment in a plane is a convex set. Moreover, it has no interior points, as any disc around any of its points contains points of the plane not on the line. Thus all its points are boundary points. Now two line segments which cross each other are convex sets with only boundary points in common, but they cannot be separated by a line. The trouble is precisely that both sets have empty interiors. This need not worry us here, as all sets we shall meet have non-empty interiors. But the problem can be serious in more advanced work, particularly in infinite dimensional cases. Thus we can use the following theorem, even though more general results exist —

Separation Theorem: If $\mathcal{A}$ and $\mathcal{B}$ are two convex sets with no interior points in common, and if at least one of the two has a non-empty interior, then we can find a non-zero row vector θ and a number d such that

$$\theta x \begin{cases} \leqslant d & \text{for all } x \text{ in } \mathcal{A} \\ \geqslant d & \text{for all } x \text{ in } \mathcal{B} \end{cases} \tag{4.4}$$

In the standard maximization problem, $d = \theta \bar{x}$ where $\bar{x}$ is the optimum choice. The separation theorem can then be paraphrased to say:

(a) $\bar{x}$ maximizes θx over all points x in $\mathcal{A}$, and
(b) $\bar{x}$ minimizes θx over all points x in $\mathcal{B}$.

This twofold result is a consequence of the assumptions of diminishing marginal rates of transformation and of substitution, and it is this result which gives their economic importance to the separate assumptions. This is because it raises the possibility of *decentralized* economic decisions. To give the simplest interpretation, interpret the problem as one of producing a bill of goods x in a one-consumer economy to maximize the utility $F(x)$ subject to the constraints $G(x) \leqslant c$. The solution yields not only $\bar{x}$, but also the equation of the common tangent. Now suppose we announce θ to be the vector of prices for the goods. Then the result (a) above says that the optimum $\bar{x}$ would be produced by an entrepreneur maximizing the value of output, while (b) says that $\bar{x}$ would also be demanded by the consumer trying to reach the utility level $F(\bar{x})$ with the minimum expenditure. If we assume away some technical complications that arise when there are free goods, this is equivalent to maximizing utility subject to the budget constraint $\theta x \leqslant d$. This separation of decisions has two advantages. One is informational: the producer need know nothing about the consumer's tastes, and the consumer need know nothing about the production technology. For each, the relevant information about the other is adequately summarized by the prices. The other relates to incentives: the process relies on the self-interest of each side to ensure the effective implementation of the optimum.

To extend this to the more meaningful case of many producers and many consumers, we need further assumptions. Specifically, externalities and income distribution problems must be either absent or efficiently resolved in the process. But even if these major restrictions are granted, a critical problem remains. Basically, the optimum quantities $\bar{x}$ and the prices θ emerge in the same calculation, and the two approaches are formally equivalent. The informational gain would be illusory if the calculation of the prices required detailed information about resources, technology and tastes, while many would regard the desirability of relying on self-interest to be dubious at best.

The issue of the relative advantages of centralized and decentralized planning is an area of very active research. One line is to calculate the information flows in the two processes; this leads to some difficult theory. Another is to ask whether workable approximations to the optimum prices can be found without solving the whole optimization problem in detail. There are special cases of some importance, such as that of a small open economy, where this is possible. However, general results are rare, and there are some very serious difficulties in letting the markets themselves find such approximations by a dynamic process. Finally, the realistic feature of uncertainty produces a difference between planning by quantities and planning through prices. These developments are matters for further reading by interested readers.

If we do not assume both $\mathcal{A}$ and $\mathcal{B}$ to be convex, full decentralization is not possible. Figure 4.2 illustrates this. In case (a) there, $\mathcal{B}$ is not convex and $\bar{x}$ does not minimize θx over it. In case (b), $\mathcal{A}$ is not convex and $\bar{x}$ does not maximize θx over it. The latter is the more common case, arising from increasing returns in production. In such a case, considerations of marginal benefits and costs have to be supplemented by an examination of the discrete choice of whether to produce at all. This leads us to look at consumers' surplus or related concepts.

We must next find conditions on the underlying functions F and G which correspond to convexity of the sets $\mathcal{B}$ and $\mathcal{A}$. Since $\mathcal{B}$ is defined as the set of points x for which $F(x) \geqslant v$, the function F should be such

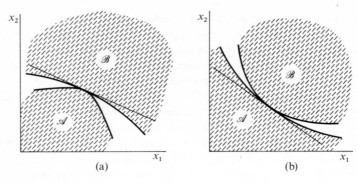

FIG. 4.2

that whenever $F(x) \geqslant v$ and $F(x') \geqslant v$ for points x and x', and δ is a number satisfying $0 \leqslant \delta \leqslant 1$, we also have $F(\delta x + (1 - \delta)x') \geqslant v$. Of course we do not know v in advance, so we should be prepared by imposing this condition for all v at the outset. A function F for which, for all points x and x' in its domain of definition, for all numbers v in its range, and for all numbers δ satisfying $0 \leqslant \delta \leqslant 1$, the inequalities $F(x) \geqslant v$ and $F(x') \geqslant v$ together imply $F(\delta x + (1 - \delta)x') \geqslant v$, will be called a *quasi-concave* function. The term may seem rather odd, but the reason for it will appear in the next chapter.

Similarly, whenever $G(x) \leqslant c$, $G(x') \leqslant c$, and $0 \leqslant \delta \leqslant 1$, we should have $G(\delta x + (1 - \delta)x') \leqslant c$ for $\mathscr{A}$ to be convex. We do know c in advance, but we are likely to try different values for it when doing comparative statics. We should therefore impose this condition for all c, and a function fulfilling it will be called *quasi-convex*. Now we can state our result in terms of the properties of the functions defining the two sets as follows —

> If $\bar{x}$ maximizes $F(x)$ subject to $G(x) \leqslant c$, where F is quasi-concave and G is quasi-convex, then there is a row vector $\theta \neq 0$ such that
> (a) $\bar{x}$ maximizes θx subject to $G(x) \leqslant c$, and
> (b) $\bar{x}$ minimizes θx subject to $F(x) \geqslant F(\bar{x})$.

The generalization to several constraints is straightforward. The set $\mathscr{A}_i$ of points for which $G^i(x) \leqslant c_i$ will be convex if G^i is quasi-convex. If this is so for all i, then the set $\mathscr{A}$ of points satisfying all the constraints, being the intersection of the convex sets $\mathscr{A}_i$, is itself convex; this is quite easy to verify from the definition of a convex set.

We can write all the constraints together in vector form as $G(x) \leqslant c$, where the inequality $\leqslant$ for vectors is simply the same inequality component by component. There are other types of vector inequalities that will be used later. The weak inequality above does not exclude the special case of equality of all components. If we wish to exclude it, so that at least one component inequality will be strict ($<$), the symbol $<$ will be used for the vector inequality. If we want all component inequalities to be strict, we shall use $\ll$ for the vector inequality. Similarly in reverse, $\geqslant$ is weak inequality in each component, $>$ strengthens this to a strict inequality in at least one component, and $\gg$ denotes strict inequality component by component.

Another advantage of using inequality constraints is that it is no longer necessary to restrict them to be fewer in number than the choice variables. The feasible set of choices can be non-trivial even with more constraints. Figure 4.3 shows some examples of this. Case (a) has two constraints, and depending on the slope of the level curves of the criterion, the optimum could be either at the corner where both constraints hold with equality, or on either face where one of them must be a strict inequality. This illustrates how it may be *desirable* to leave some resource unused. In case (b) with three constraints, there is in general no point where all three hold with equality, and it becomes *necessary* to leave at least one of them not binding. Which one is left as

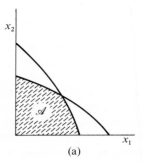

(a)

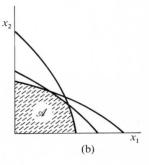

(b)

FIG. 4.3

a strict inequality depends on the criterion function. If the optimum is on one of the three faces, then two constraints will not be binding. In the case of linear programming, where F and G are linear functions, it is possible to make more precise statements about the number of binding constraints.

Throughout this discussion we have only required $\mathscr{A}$ and $\mathscr{B}$ to be convex. Their boundaries need not be smooth curves, and can have kinks or flat segments. This raises a number of possibilities, some of which are shown in Figure 4.4. In case (a), two corners happen to meet at the optimum. Now we can find many lines through $\bar{x}$ which separate the two sets, i.e. θ is not unique. None of these lines can be called a

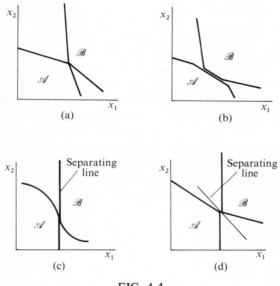

FIG. 4.4

common tangent in the usual sense, but that is not essential for the economics of the problem. Decentralization depends only on the separation property, namely that the two sets lie one on each side of the line $\theta x = d$. Thus separation is a generalization of the notion of a common tangent, and that is how we dispense with the requirements of differentiability of F and G. In case (b), the two sets have a flat portion in common. This need not worry us unduly, for all candidates for the optimum choice along this common segment must have the same value of $F(x)$, and that, after all, is the magnitude that interests us. There is, however, a problem about decentralization. Given θ, all points on the flat portion of $\mathscr{A}$ will yield equal value of output to the producer, and all those on the flat portion of $\mathscr{B}$ will yield the same utility to the consumer. Their choices will be arbitrary to that extent, and there is no reason why the independent choices should coincide. We can only make a weaker claim, namely that if the two happen to make coincident choices, neither will have any positive incentive to depart from these choices. This is a standard procedure in any careful statement of economic equilibrium theory.

If the two boundaries have vertical parts at the optimum, we may have a vertical separating line, corresponding to $\theta_2 = 0$. This is the case in (c). However, in case (d) it is also possible to have non-vertical separating lines even though the boundaries have vertical parts at the optimum. Similarly for horizontal parts leading to the possibility of $\theta_1 = 0$. This shows that without stronger assumptions, it is not possible to guarantee strictly positive prices. In fact, if the boundaries sloped upward at the optimum, the common tangent would have a positive slope, and one of the prices would be negative. This is usually avoided by assuming either (a) there is free disposability of both goods, when the boundary of $\mathscr{A}$ cannot slope upward, or (b) both goods are desirable, so that the boundary of $\mathscr{B}$ cannot slope upward. Both these assumptions have been implicit in all the illustrative figures.

Finally, we should note that nothing of economic substance will change if we multiply the row vector θ and the related numbers like d by the same positive number. Another way of saying the same thing is that only the relative prices like θ_1/θ_2 matter. Of course, these relative prices equal the common value of the appropriate marginal rates of transformation and substitution at the optimum when we have smooth curves, and provide the appropriate generalizations in terms of the notion of separability otherwise.

This chapter has introduced some basic mathematical concepts for handling inequality constraints, and carried the analysis to the point of defining and interpreting 'prices' associated with the outputs or the choice variables themselves. In the next chapter, these concepts will be used for obtaining the shadow prices associated with the resource constraints.

EXAMPLES

Example 4.1   To illustrate separation, consider a simple case

$$F(x, y) = xy \qquad \text{and} \qquad G(x, y) = x^2 + y^2.$$

Restrict x and y to non-negative values, and consider the set $\mathscr{A}$ defined by $G(x, y) \leqslant 25$ and the set $\mathscr{B}$ defined by $F(x, y) \geqslant v$ for various values of v. It is well known that $\mathscr{A}$ is a quarter disc, and $\mathscr{B}$ a rectangular hyperbola and points above. Each set is convex.

Figure 4.5 illustrates this. For $v = 10$, the two sets have interior

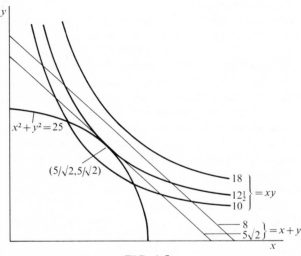

FIG. 4.5

points in common and cannot be separated. For $v = 12.5$, they have only the boundary point $(5/\sqrt{2}, 5/\sqrt{2})$ in common, and we can separate them by choosing $\theta_1 = 1$, $\theta_2 = 1$, and $d = 5\sqrt{2}$, or any positive multiple of all three numbers. For even larger values of v, e.g. $v = 18$, the sets do not have any points in common. We can then separate them *strictly*, i.e. find a θ and d such that the inequalities in (4.4) hold strictly. An example would be to take $\theta_1 = 1$, $\theta_2 = 1$ and $d = 8$.

Example 4.2   To illustrate the importance of inequality constraints in another context, consider a consumer with a utility function $ax + \log y$. In the familiar notation, mechanical application of the conditions gives

$$a = \pi p \qquad \text{and} \qquad 1/y = \pi q.$$

Using the budget constraint,

$$m = px + qy = (ax + 1)/\pi.$$

So

$$a = p(ax + 1)/m,$$

and hence the demand functions

$$x = m/p - 1/a, \qquad y = p/(aq).$$

If $m < p/a$, the demand for x becomes negative. This may be possible in some cases, e.g. in a portfolio selection problem where 'short sales' are allowed. Generally, however, we will require such quantities to be non-negative, and the only way to ensure that is to incorporate an explicit constraint $x \geqslant 0$ in the problem.

Example 4.3   For yet another illustration of inequality constraints, consider the problem of distributing income between two consumers who envy each other. If the first is given an income of y_1 and the second y_2, their utilities are respectively

$$u_1 = y_1 - ay_2{}^2 \qquad \text{and} \qquad u_2 = y_2 - ay_1{}^2,$$

where a is a positive constant; thus each derives disutility from income given to the other. Suppose the criterion of social welfare admits such feelings of envy, and simply maximizes the sum of utilities, $u_1 + u_2$.

Even if there were no constraints on the aggregate income available, this maximization problem would have a finite solution. It is easy to verify that the unconstrained maximum is attained when $y_1 = y_2 = 1/(2a)$. Therefore, even if aggregate income in excess of $(1/a)$ were available, we would choose not to use it. The envy effects would become so strong as to overwhelm the additional utility each consumer would obtain from his own additional income. In view of this, if y^* is the aggregate income given, the constraint should be expressed as $y_1 + y_2 \leqslant y^*$, and whether or not the constraint holds as an equality should be answered in the process of solution of the problem.

EXERCISES

4.1   How would you adapt the concepts and analyses of this chapter in order to handle constrained minimization problems with inequality constraints?

4.2   If we made assumptions which rule out the possibility of the boundaries of the sets $\mathscr{A}$ and $\mathscr{B}$ having flat segments, the optimum choice would be unique. Examine how the definitions of quasi-concavity and quasi-convexity need to be strengthened in order to achieve this.

4.3   How is Figure 4.1 altered when (a) one of the choice variables is labour, which gives disutility to consumers and is an input to production, and (b) when one of them is pollution, which gives disutility to consumers and is a by-product of production of a good which is the other choice variable? Interpret the associated 'prices' in each of these cases.

FURTHER READING

For an excellent discussion of separation theorems and the economics of decentralization, see Koopmans, op. cit (p. 23). A microeconomics textbook which uses such geometric methods as well as calculus ones is Malinvaud, op. cit (p. 23).

A detailed discussion of the various aspects of decentralization can be found in Heal, op. cit (p. 23), Section 3.3. A pioneering analysis of the implications of uncertainty for the relative desirability of price and quantity control is

WEITZMAN, M. L. 'Prices vs. Quantities', *Review of Economic Studies,* XLI(4), October 1974, pp. 477–91

The issue of optimum production decisions subject to economies of scale is discussed at an elementary level by Samuelson, op. cit. (p. 11), p. 637, and at a more advanced level by Malinvaud, op. cit (p. 23), pp. 225–9.

5. Concave Programming

The analysis of Chapter 2 shows that the Lagrange multipliers measure the trade-off between the constraints and the value of the objective. To extend this to the case of inequality constraints, we must examine such a trade-off in this context, and express it in the language of analytic geometry.

As in other situations that have to do with prices, problems arise if this trade-off shows increasing returns. To avoid this, at least to begin with, I shall place stronger restrictions on F and G than were used in the previous chapter. To draw a parallel with consumer theory, the assumption for F will be that it shows diminishing marginal utility and not merely a diminishing marginal rate of substitution.

For a function of a scalar variable, the condition of diminishing marginal utility would be a negative second derivative. This can be extended to functions of vector variables using matrices, but geometric reasoning enables us to avoid that for a long time. We can characterize such a function in terms of geometry by saying that the chord joining any two points on its graph lies entirely below the graph between the same two points. Algebraically, this can be expressed as

$$F(\delta x + (1 - \delta)x') \geq \delta F(x) + (1 - \delta)F(x') \qquad (5.1)$$

for all x, x' in the domain of F, and for all numbers δ with $0 \leq \delta \leq 1$. A function which has this property is called *concave*. This allows the special case of a straight line, and it could be excluded by requiring the inequality to be strict for $0 < \delta < 1$; such a function would be called *strictly concave*.

Concavity is a stronger requirement than quasi-concavity, i.e. every concave function is quasi-concave but not vice versa. This is in fact the reason for the term quasi-concave, which must otherwise seem rather strange.

Let F be a concave function, and suppose that each of $F(x)$ and $F(x')$ is $\geq v$ for some scalar v. Then, using (5.1), we have

$$F(\delta x + (1 - \delta)x') \geq \delta v + (1 - \delta)v = v,$$

which proves that concavity implies quasi-concavity. To show that the converse is not true, we need only remember the difference between diminishing marginal utility and diminishing marginal rate of substitution; for example, it is easy to verify that $F(x_1, x_2) = x_1 x_2$ is quasi-concave but not concave.

Two other equivalent characterizations of concave functions will be useful later. First, the set of points on and under the graph of such a function is a convex set, i.e. if (x, y) and (x', y') are such that $y \leqslant F(x)$ and $y' \leqslant F(x')$, then for $0 \leqslant \delta \leqslant 1$, we have

$$\delta y + (1 - \delta)y' \leqslant F(\delta x + (1 - \delta)x').$$

This follows at once from (5.1). Next, write that inequality as

$$[F(x' + \delta(x - x')) - F(x')]/\delta \geqslant F(x) - F(x').$$

Now let δ tend to zero. Provided F is differentiable, the chain rule shows that the left hand side tends to $F_x(x')(x - x')$, which is the linear approximation to F using its tangent at x' to approximate the curve. Thus we have

$$F(x) - F(x') \leqslant F_x(x')(x - x'). \tag{5.2}$$

In words, the change in a concave function is overestimated by its tangent at any point, i.e. any tangent to the curve lies above it. For a function of a scalar variable, it is easy to see the equivalence of these characterizations. We shall soon meet a natural generalization of (5.2) for functions that are not differentiable.

Similarly, a function G is called *convex* if, for all x, x' in its domain and for all numbers δ with $0 \leqslant \delta \leqslant 1$, we have

$$G(\delta x + (1 - \delta)x') \leqslant \delta G(x) + (1 - \delta)G(x') \tag{5.3}$$

and *strictly convex* if the inequality is strict when $0 < \delta < 1$. The set of points on and above the graph of a convex function will be a convex set, and changes in such a function will be underestimated by a linear approximation. A vector function will be convex if each of its component functions is convex. In this chapter, I assume that the criterion function is concave and the vector constraint function is convex; this is *concave programming*.

We are now ready to discuss Lagrange multipliers. Throughout the argument, I shall use the production example, with x as output levels and c as resource availabilities, for illustration and concreteness. No special interpretation will be placed on the criterion, and I shall refer to its 'value' in general terms.

Consider the problem in standard form: to maximize $F(x)$ subject to $G(x) \leq c$. The maximum value is a function of c; write it in the usual notation as $V(c)$. This is just the function which shows the trade-off between resources and value, and is therefore the crucial concept in the argument. It is tempting to identify its partial derivatives as the Lagrange multipliers at once, but we have to proceed more slowly in order to sort out some problems along the way.

The important general result on which the subsequent argument hinges is that if F is concave and G is convex, then V is concave. The proof is a mechanical verification, but this type of argument appears very frequently, and its steps are not without economic interest. It is therefore advisable to follow it carefully.

Let c and c' be any two resource endowments, and suppose that the corresponding values $v = V(c)$ and $v' = V(c')$ are attained at $\bar{x}$ and $\bar{x}'$ respectively. Since the optimum choices must be feasible, $G(\bar{x}) \leq c$ and $G(\bar{x}') \leq c'$. Now let δ be any number satisfying $0 \leq \delta \leq 1$, and ask whether it is possible to do at least as well as $\delta V(c) + (1 - \delta)V(c')$ when the resources are $\delta c + (1 - \delta)c'$, which would prove concavity of V. A natural candidate for the output vector to try is $\delta\bar{x} + (1 - \delta)\bar{x}'$. The first point to check is whether it is feasible. By the convexity of G, we have

$$G(\delta\bar{x} + (1 - \delta)\bar{x}') \leq \delta G(\bar{x}) + (1 - \delta)G(\bar{x}') \leq \delta c + (1 - \delta)c',$$

proving feasibility. The next point is to find its value. Using the concavity of F, we have

$$F(\delta\bar{x} + (1 - \delta)\bar{x}') \geq \delta F(\bar{x}) + (1 - \delta)F(\bar{x}') = \delta V(c) + (1 - \delta)V(c').$$

Since we have found a feasible vector yielding value at least as high as the expression on the extreme right, the maximum value, $V(\delta c + (1 - \delta)c')$, can be no smaller. This is the result we are trying to prove.

The economics behind this is that the convexity of G rules out increasing returns, thus ensuring that a weighted average of outputs can be produced given the same weighted averages of resources, and then the concavity of F results in its yielding at least the same weighted average of values.

As V is a concave function, the set of points on or below its graph is a convex set. This set $\mathscr{A}$ is the collection of points (c, v) such that $v \leqslant V(c)$, i.e. value of at least v can be produced using resources of no more than c. Therefore it is natural to think of it as the set of production possibilities for 'value'. Clearly, given any point in $\mathscr{A}$, all points to the south-east of it are also in $\mathscr{A}$. (Equivalently, V is a monotonic non-decreasing function.) This is because we have written the constraints so that an increase in c widens the choice. The set is $(m + 1)$ dimensional when m is the number of constraints. Figure 5.1 shows it for the case

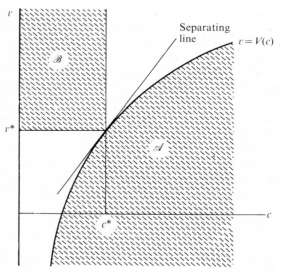

FIG. 5.1

$m = 1$. We see that V being an increasing concave function corresponds to a positive but diminishing marginal return to the resource in producing value.

Convex sets are meant to be separated from other convex sets. To do this in the most useful way, choose a point (c^*, v^*) in $\mathscr{A}$ such that $v^* = V(c^*)$. This must be a boundary point, since the point $(c^*, v^* - r)$ is in $\mathscr{A}$ and $(c^*, v^* + r)$ is not in $\mathscr{A}$, for any positive r. Now let $\mathscr{B}$ be the set of all points (c, v) such that $c \leqslant c^*$ and $v \geqslant v^*$, i.e. value v cannot be attained with resources c save when $c = c^*$ and $v = v^*$. Thus the set $\mathscr{B}$ serves the same function as the corresponding set in Chapter 4. Clearly $\mathscr{B}$ is a convex set with a non-empty interior, and $\mathscr{A}$ and $\mathscr{B}$ have only boundary points in common, thus the separation theorem can be applied. For reasons that will become clear in a moment, I write the equation of the separating hyperplane as

$$\omega - \pi c = d = \omega^* - \pi c^*$$

with the signs arranged so that

$$\omega - \pi c \begin{cases} \leqslant d & \text{for all } (c, v) \text{ in } \mathscr{A} \\ \geqslant d & \text{for all } (c, v) \text{ in } \mathscr{B} \end{cases} \tag{5.4}$$

The first point to note is that the number ι and the row vector π must both be non-negative. For example, suppose that ι is negative. Now consider the point $(c^*, v^* + 1)$, which is clearly in $\mathscr{B}$. We have

$$\iota(v^* + 1) - \pi c^* < \omega^* - \pi c^* = d,$$

which contradicts the separation property. Similarly, considering points $(c^* - e_i, v^*)$ where e_i is a vector with its ith component equal to 1 and all other components zero, we find that π_i must be non-negative, for each i.

Next observe that (c^*, v^*) maximizes $(\omega - \pi c)$ over $\mathscr{A}$. This has an important interpretation. Consider a hypothetical producer who 'manufactures' value of the criterion out of the inputs. If he is paid a price ι for each unit of value, and charged prices π for use of the inputs c, then a production plan (c, v) will yield him a profit of $(\omega - \pi c)$. Then (c^*, v^*) will be a profit-maximizing choice for him from among all conceivable plans, i.e. the whole set $\mathscr{A}$. There may be an aggregate constraint of c^* on resource availability, but there is no need for the producer to be aware of it, since he will not wish to violate it anyway The interpretation is special, but the principle is general and important: constrained choice can be converted into unconstrained choice if the proper scarcity costs or shadow values of the constraints are netted out of the criterion function. To the economist, this is the most important feature of Lagrange's method in concave programming.

Again, only relative prices matter, and nothing of any substance is changed if we multiply ι, π, and d by any positive number. This raises an attractive possibility: if we choose marginal value itself as numéraire, thus setting $\iota = 1$, then the resource prices π will become precise generalizations of the Lagrange multipliers of Chapter 2. But before we choose a numéraire, we must ensure that it is not a free good, and nothing so far guarantees $\iota > 0$. The entire vector (ι, π) cannot be zero, but that is not enough.

Let us see what happens if $\iota = 0$. Then at least one component of π must be non-zero, i.e. positive. The equation of the separating hyperplane becomes $-\pi c = -\pi c^*$, i.e. $\pi(c - c^*) = 0$. For all (c, v) in $\mathscr{A}$, we have $-\pi c \leqslant -\pi c^*$, i.e. $\pi(c - c^*) \geqslant 0$. In the one constraint case this means that the hyperplane is vertical at c^*, and the entire set $\mathscr{A}$ lies to its right. This means that production is impossible at a level in the domain of definition of F if the resource availability is less than c^*. This is commonly caused by indivisibilities.

Figure 5.2 shows two examples of this. In case (a), the marginal

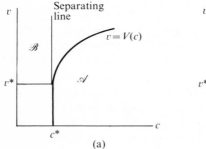
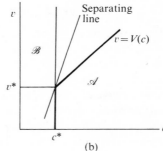

FIG. 5.2

product of the resource is infinite at c^* and falls gradually; thus only a vertical separating line will do. In case (b) this is not so, and while a vertical separating line exists, it is also possible to find such lines of finite slope, and thus positive ι. This shows that the conditions soon to be found for ensuring this are only sufficient and not necessary.

The concept of indivisibility gives us a hint for finding a natural condition. If the set $\mathscr{A}$ has any points to the left of c^*, its boundary cannot have an infinite slope at (c^*, v^*). For this to be true, there must be an x^0 such that $G(x^0) < c$ in the domain of definition of F, for then we can choose $(G(x^0), F(x^0))$ as the desired point. If there are several constraints, we must assume this for each of them, i.e. that there is an x^0 such that $G(x^0) \leqslant c$. This condition will be called the *constraint qualification*. It is possible to use a much weaker condition, and thereby have a stronger result. But the proof is quite complicated, and is best left for more advanced work.

It is easy to prove formally that the constraint qualification ensures a positive ι. Otherwise at least one component of π would be positive. Now every component of $(G(x^0) - c)$ is negative. So if we multiply the corresponding components of these vectors, we will have all non-positive products with at least one actually negative. Adding them together gives $\pi(G(x^0) - c) < 0$. However, the point $(G(x^0), F(x^0))$ is in $\mathcal{A}$, and by the separation property we have $-\pi G(x^0) \leqslant -\pi c$, i.e. $\pi(G(x^0) - c) \geqslant 0$. This contradiction forces us to conclude that the supposition $\iota = 0$ must be wrong, thus proving the result.

Henceforth I shall assume the constraint qualification to be satisfied, and normalize to $\iota = 1$.

Now for any c, the point $(c, V(c))$ is in $\mathcal{A}$. So by the separation property we have $V(c) - \pi c \leqslant V(c^*) - \pi c^*$, or

$$V(c) - V(c^*) \leqslant \pi(c - c^*). \tag{5.5}$$

The linear function on the right hand side thus overestimates changes in V. This looks very much like (5.2), thus strengthening our idea that π is closely related to $V_c(c^*)$, the vector of partial derivatives of V at the initial point c^*. But one difficulty remains: we cannot be sure that V is differentiable. So far in this chapter we have not even assumed F and G to be differentiable, but even if they are, V may fail to be. This is because different inequality constraints may hold as exact equalities for different values of the parameters, and in the process of moving from one such régime to another the slope of V may change suddenly. Consider a case where some resource is just on the point of becoming superfluous at the margin. Any further increment in it will be left unused, and the 'rightward' partial derivative of V will be zero. What happens for a slight reduction in the amount available depends on whether the point of superfluity is reached with the marginal product of the resource dropping smoothly. If so, a small decrease in its availability will cause a second order small loss in value, and the 'leftward' partial derivative will be zero as well. If the marginal product stays above a positive level before reaching this point, then the leftward partial derivative will have to be positive, and any multiplier between this value and zero will do for separation. This is the case in linear programming, where the marginal product is constant because of linearity right up to the point where the constraint ceases to bind.

Even when such discontinuities exist, a very natural generalization of the concept of diminishing returns holds. The leftward partial is never less than the rightward, which is like saying that the marginal product of the kth dose of a resource cannot exceed that of the $(k-1)^{\text{th}}$. This is a simple consequence of the concavity of V, which is really the economically important property.

The asterisks having served their purpose of distinguishing a particular point in the (c, v) space for separation, let us drop them, and consider a point $(c, V(c))$ with its associated multipliers π, and compare it with a neighbouring point $(c + he_i, V(c + he_i))$, where h is a number and e_i a vector with its ith component equal to 1 and all others zero. As in (5.5) we have

$$V(c + he_i) - V(c) \leqslant h\pi_i.$$

If h is positive, we can divide by it to write

$$[V(c + he_i) - V(c)]/h \leqslant \pi_i.$$

It is easy to show that as V is a concave function, the left hand side is a monotonic non-increasing function of h, and therefore must have a limit as h goes to zero. This limit is what I have been calling the 'rightward' partial derivative, which I shall denote by $V_i(c)_+$. Thus we have proved that $V_i(c)_+ \leqslant \pi_i$. If h is negative, division reverses the direction of the inequality, and defining the leftward partial $V_i(c)_-$ similarly, we have $V_i(c)_- \geqslant \pi_i$. Thus we have the final result generalizing the notion of diminishing returns and relating the multipliers to these derivatives:

$$V_i(c)_- \geqslant \pi_i \geqslant V_i(c)_+ \tag{5.6}$$

This chapter has built up the desired interpretation of the multipliers in terms of the maximum value function. The next chapter will complete the story by considering the implications in terms of the choice variables x. Then the relevant results can be stated precisely, and some applications discussed.

EXAMPLES

Example 5.1   To illustrate the constraint qualification, we have the famous problem of maximizing $F(x, y) = xy$ subject to $G(x, y) = (x + y - 1)^3 \leqslant 0$. The constraint will turn out to be binding, and we can write the conditions in terms of a multiplier π as

$$y - 3\pi(x + y - 1)^2 = 0$$
$$x - 3\pi(x + y - 1)^2 = 0.$$

But the constraint is $(x + y - 1)^3 = 0$, so $(x + y - 1)^2 = 0$ and therefore the conditions become $x = y = 0$. However, this violates the constraint.

Conversely, suppose we use the conditions to derive $x = y$, and then use the constraint to conclude that $x = y = \frac{1}{2}$. This will in fact turn out to be the correct solution. However, each condition then becomes $\frac{1}{2} - \pi 0 = 0$. This can be true only if π is infinite.

Since only relative prices matter, an infinite π is equivalent to a zero ι in our earlier notation. Thus the constraint qualification must have failed. Unfortunately we cannot check this directly since the form we used works only for convex G.

However, we can relate the problem to the condition in Chapter 1 which required at least one partial derivative of G to be non-zero at the optimum. In this case each of these is $3(x + y - 1)^2$, so both are zero when $x = y = \frac{1}{2}$. Then, recalling the definition of π as the common value in eqn. (1.5), we see why it is infinite in this case.

Example 5.2   Consider the maximization of

$$F(x) = 1 + [1 - (x - 2)^2]^{\frac{1}{2}}$$

subject to $x \leqslant c$, where the positive value of the square root is taken. The graph of the function in (x, y) space is then the upper semi-circle of a circle of radius 1 and centre $(2,1)$. The function is defined only for $1 \leqslant x \leqslant 3$. This is shown in Figure 5.3, which also shows the corresponding maximum value function $v = V(c)$. For $c < 1$ the function is not defined. For $1 \leqslant c \leqslant 2$ it follows the function $F(x)$. However, when $c > 2$, it becomes desirable to maintain $x = 2$, thus achieving the value $V(c) = 2$ when $F(c)$ would be smaller, or undefined for $c > 3$. Thus $V(c)$ remains constant at 2 for $c \geqslant 2$.

This example illustrates another constraint qualification problem, as well as the need for inequality constraints.

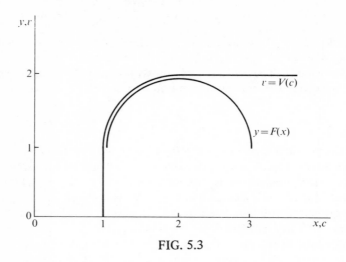

FIG. 5.3

5.1   Reformulate the analysis of this chapter, including the appropriate concavity and convexity conditions to be imposed on the various functions, to deal with constrained minimization problems. Draw the analogue of Figure 5.1, and obtain the multipliers from a separation argument.

5.2   For the problem of maximizing $F(x) = \frac{1}{2}x + \sin x$ subject to $x \leqslant c$, draw the maximum value function $v = V(c)$.

(Note: For drawing the graph of $F(x)$, recall that $d(\sin x)/dx = \cos x$, and that $\cos x < -\frac{1}{2}$ when x is between $120°$ and $240°$, and of course periodically at $360°$ intervals.)

A more general treatment of the constraint qualification can be found in GALE, D. 'A Geometric Duality Theorem with Applications', *Review of Economic Studies*, XXXIV(1), January 1967, (pp. 19–24).

For a proper mathematical treatment of convex sets etc. see EGGLESTON, H. G. *Convexity*, Cambridge University Press, Cambridge, 1963.

6. Results and Applications

To complete the discussion of concave programming, let us recast the discussion surrounding Figure 5.1 in terms of the underlying choice variables. Suppose $\bar{x}$ maximizes $F(x)$ subject to $G(x) \leq c$, and let π be the vector derived from the separating hyperplane at $(c, V(c))$; remember that we have dropped the asterisks. Now the point $(F(\bar{x}), G(\bar{x}))$ is in $\mathscr{A}$, and from the separation property (5.4) we have

$$F(\bar{x}) - \pi G(\bar{x}) \leq V(c) - \pi c. \tag{6.1}$$

Of course $F(\bar{x}) = V(c)$, and therefore

$$\pi[c - G(\bar{x})] \leq 0. \tag{6.2}$$

This causes a problem. Every component of π is non-negative, and since $\bar{x}$ satisfies the constraints, every component of $[c - G(\bar{x})]$ is also non-negative. So every term in the inner product of these vectors on the left hand side of (6.2) is non-negative. There is only one way in which the sum of such terms could be non-positive and that is for each of these terms and therefore the whole inner product to be zero. Thus, for each i, $\pi_i[c_i - G^i(\bar{x})] = 0$, i.e. at least one of these two factors must be zero. The whole result can be stated in the form that for each i, we must have

$$\pi_i \geq 0, G^i(\bar{x}) \leq c_i \qquad \text{with at least one equality.} \tag{6.3}$$

Then both (6.1) and (6.2) also become equalities.

This is the important economic implication of inequality constraints that was mentioned in Chapter 2, for (6.3) says that each resource is either fully used or has a zero shadow price. Note that there is nothing to prevent both $\pi_i = 0$ and $G^i(\bar{x}) = c_i$ being true for any i. This can happen when a constraint is just about to cease being binding. What (6.3) rules out is the possibility of an unused resource having a positive shadow price.

When two vector inequalities are such that in each component pair at least one exact equality must hold, i.e. no two component inequalities can be slack together, we say that the vector inequalities show *complementary slackness*. Thus we can restate (6.3) as

$$\pi \geq 0, G(\bar{x}) \leq c \qquad \text{with complementary slackness.} \tag{6.3}$$

Next consider the fact that for any x, the point $(F(x), G(x))$ is in the set $\mathscr{A}$. Since (6.1) has been proved to be an equality, the separation property can be written as

$$F(x) - \pi G(x) \leqslant F(\bar{x}) - \pi G(\bar{x}), \qquad (6.4)$$

i.e. $\bar{x}$ maximizes $F(x) - \pi G(x)$ without any constraints. This is an alternative statement in terms of the underlying functions of what Lagrange's method achieves, and is more convenient than the earlier statement in terms of the set $\mathscr{A}$. This completes the characterization stated in the result —

> Suppose F is a concave function and G a vector convex function, and that there exists an x^0 satisfying $G(x^0) \ll c$. If $\bar{x}$ maximizes $F(x)$ subject to $G(x) \leqslant c$, then there is a row vector $\pi \geqslant 0$ such that
> (i) $\bar{x}$ maximizes $F(x) - \pi G(x)$ without constraints, and
> (ii) $\pi \geqslant 0, G(\bar{x}) \leqslant c$ show complementary slackness.

None of this requires F and G to have derivatives. If they do happen to be differentiable, the first-order conditions necessary for (i) above are

$$F_x(\bar{x}) - \pi G_x(\bar{x}) = 0. \qquad (6.5)$$

This looks exactly like (1.10), but the inequality constraints make a difference. To solve for $\bar{x}$ and π, we must now use (6.5) together with the complementary slackness conditions (6.3). Each pair of these contributes one equation, and there is no difficulty in principle about having enough equations. But we do not know in advance for any i whether that equation is going to be $\pi_i = 0$ or $G^i(\bar{x}) = c_i$. We may have to resort to the crude device of trying all possible combinations (2^m of them) and checking each for consistency, hoping to rule out all but one. This can be very tedious, but that is a price to be paid for the economic realism of inequality constraints. After a little experience, we can tell for many standard economic problems which constraints are sure to hold as equalities, and this cuts down the number of cases to be checked. Thus, for a consumer who is not satiated, we can be sure that his budget constraint will be binding.

If F and G are not differentiable, we can establish inequalities for the leftward and rightward derivatives using techniques now familiar:

$$F_x(\bar{x})_- - \pi G_x(\bar{x})_- \geq 0 \geq F_x(\bar{x})_+ - \pi G_x(\bar{x})_+ \qquad (6.6)$$

The solution is then even more complicated.

There is another point where the problem of this chapter differs from that in Chapter 1. The conditions there were derived without any reference to the concavity or convexity of functions. It is possible to use separation arguments to obtain such necessary conditions without assuming concavity or convexity even with inequality constraints. This involves some rather specialized theorems in mathematics, and I shall not go into the subject here, but merely mention some important differences involved. First, conditions (6.5) remain valid, but exactly the same conditions would result for a problem of minimizing $F(x)$, or of maximizing it with respect to some variables and minimizing it with respect to others, or in general for a *stationary point* of F. Further, the same conditions apply to a *local* stationary point, i.e. where F is stationary in comparison with points in some small neighbourhood. Thus the first-order conditions are not sufficient for a true *global* maximum. In the case of concave programming, we shall soon see that they are.

Secondly, a different constraint qualification is necessary. Finally, and most important, even if $\bar{x}$ is the true global maximizing choice, without concavity we cannot be sure that it maximizes the Lagrangean; it may merely yield a stationary point of it. The problem is similar to that of determining the optimum output when there are economies of scale. The first-order condition of equality between price and marginal cost is still necessary, but profit need not be maximized even locally. The interpretation of Lagrange's method as converting constrained 'value' maximization to unconstrained 'profit' maximization must be confined to the case of concave programming.

However, if we find an $\bar{x}$ that maximizes the Lagrangean expression and shows complementary slackness, then we can be assured that it is a global maximizing choice. This yields sufficient conditions. To prove the result, consider any feasible x, i.e. one satisfying $G(x) \leq c$. Since $\bar{x}$ maximizes L without any constraints, we have, *a fortiori*,

$$F(\bar{x}) - \pi G(\bar{x}) \geq F(x) - \pi G(x).$$

Next, remember that $\pi_i \geq 0$ for each i. If we multiply $G^i(x) \leq c_i$ by π_i and add, we will find $\pi G(x) \leq \pi c$ for the matrix products. However, if this is done for $\bar{x}$, we will have either $G^i(\bar{x}) = c_i$ or $\pi_i = 0$, and therefore $\pi_i G^i(\bar{x}) = \pi_i c_i$ for each i. Adding, $\pi G(\bar{x}) = \pi c$. The two together yield

$$\pi G(\bar{x}) \geqslant \pi G(x),$$

and adding,

$$F(\bar{x}) \geqslant F(x).$$

Since x could have been any feasible choice, we have proved that $\bar{x}$ is a global maximizing choice. The argument so far has not used concavity at all. The need for it arises because an $\bar{x}$ satisfying (i) is not easy to find in the abstract. If $(F - \pi G)$ is concave, for which it is sufficient to have F concave and G convex, then the task is simplified. We need only find an $\bar{x}$ satisfying (6.5), or in the absence of differentiability (6.6), and it will do the job. In the differentiable case, for example, knowing that a linear approximation overestimates changes in a concave function, we have

$$[F(x) - \pi G(x)] - [F(\bar{x}) - \pi G(\bar{x})] \leqslant [F_x(\bar{x}) - \pi G_x(\bar{x})] \ (x - \bar{x}) = 0.$$

In the more general case, the same result follows from separate linear approximations to the right and the left. All this is summed up as follows –

> If $\bar{x}$ and π are such that
> (i) $\bar{x}$ maximizes $F(x) - \pi G(x)$, and
> (ii) $\pi \geqslant 0$, $G(\bar{x}) \leqslant c$ show complementary slackness,
> then $\bar{x}$ maximizes $F(x)$ subject to $G(x) \leqslant c$. If $(F - \pi G)$ is a
> concave function, or even more strongly, F is a concave
> function and G a convex function, then (6.5) or (6.6) will be
> sufficient for (i) above.

Note that no constraint qualification appears in the sufficient conditions.

In many economic problems, a natural requirement is that the choice variables should be non-negative. It may be optimum in some cases to meet some of these constraints with equality: specialization of production in some cases of international trade is an instance of this. We can use the results above to take care of such constraints quite easily, since $x \geqslant 0$ can be written as $-x \leqslant 0$, and $-x$ is a convex function. But the special case is of such frequent occurrence that it will be useful to state the form of the result explicitly for it. Suppose we have some constraints $G(x) \leqslant c$ in addition to the non-negativity ones. All we have to do is to define a vector of multipliers π for these other constraints and another, ρ, for the constraints $-x \leqslant 0$, and the result for necessary conditions becomes —

> Suppose F is a concave function and G a vector convex function, and that there exists an x^0 satisfying $G(x^0) \ll c$, $x^0 \geqslant 0$. If $\bar{x}$ maximizes $F(x)$ subject to $G(x) \leqslant c$, $x \geqslant 0$, then there are row vectors π and ρ of appropriate dimensions such that
>
> (i) $\bar{x}$ maximizes $F(x) - \pi G(x) + \rho x$ without constraints,
> (ii) $\pi \geqslant 0$, $G(\bar{x}) \leqslant c$ show complementary slackness, and
> (iii) $\rho \geqslant 0$, $\bar{x} \geqslant 0$ show complementary slackness.
> If F and G are differentiable, then (i) implies

$$F_x(\bar{x}) - \pi G_x(\bar{x}) + \rho = 0. \tag{6.7}$$

Otherwise we have the appropriate left and right inequalities. These, and a statement of similar sufficient conditions, are left as exercises.

EXAMPLES

Example 6.1   The simplest illustration of the effect of non-negativity conditions is that of maximizing $F(x)$ for a scalar variable x, with $x \geqslant 0$ as the only constraint. Then (6.7) becomes $F'(\bar{x}) = -\rho \leqslant 0$, and at least one of ρ and $\bar{x}$ must be zero by complementary slackness. Thus we have two possibilities, either $\bar{x} = 0$ with $F'(0) \leqslant 0$, or $\bar{x} > 0$ with $F'(\bar{x}) = 0$. A simple sketch will show the meaning of this. It will also show how the same conditions are sufficient if F is concave.

Example 6.2   The methods of this chapter enable us to solve the problem introduced in Example 4.2. Suppose we are to maximize $F(x, y) = ax + \log y$, subject to the constraints $px + qy \leqslant m$, $x \geqslant 0$ and $y \geqslant 0$, To save space, I shall assume it known that the budget constraint must hold as an equality, thus removing the need for some checking, and that the constraint $y \geqslant 0$ will not bind; then its multiplier will be zero and there will be no need to introduce it at all. However, as we saw, we do not know in advance whether $x \geqslant 0$ will matter. Let ρ be the multiplier for it, and π that for the budget constraint. Then our conditions are

$$a - \pi p + \rho = 0, \; 1/\bar{y} - \pi q = 0, \qquad \text{and}$$

$$\rho \geqslant 0, \bar{x} \geqslant 0 \quad \text{with complementary slackness.}$$

Let us try the different possibilities.

First suppose $\rho > 0$. Then by complementary slackness $\bar{x} = 0$, and from the budget constraint $\bar{y} = m/q$. The second of the derivative conditions implies $\pi = 1/(q\bar{y}) = 1/m$, and finally from the first,

$$\rho = \pi p - a = p/m - a.$$

This is consistent provided $p/m - a > 0$.

Next suppose $\bar{x} > 0$. Using complementary slackness, the conditions become those of Example 4.2, and tracing the steps there, we have consistency provided $p/m - a < 0$.

Finally, we can have both $\bar{x}$ and ρ zero if $p/m = a$. These three cases are mutually exclusive and exhaustive, i.e. one and only one of them must hold, and therefore the solution is complete.

Example 6.3   The most important application of the results of this chapter is the theory of *linear programming*. Here we try to maximize a linear function

$$F(x) = ax \tag{6.8}$$

subject to linear constraints and non-negativity constraints

$$G^i(x) = b^i x \leqslant c_i \qquad \text{for} \qquad i = 1, 2, \ldots m$$

$$x \geqslant 0, \tag{6.9}$$

where a and b^i are n-dimensional row vectors. Stacking the b^i vertically into an m-by-n matrix B, we can write the constraints in vector form

$$G(x) = Bx \leqslant c. \tag{6.10}$$

The concavity and convexity conditions are fulfilled. So is the constraint qualification if the constraints do not reduce the feasible choices to a space of dimension smaller than n. As a matter of fact, for the kinds of reasons that Figure 5.2 explained, this will not matter anyway.

We have the partial derivatives $F_j(x) = a_j$ and $G_j{}^i(x) = b_j{}^i$. therefore $F_x(x) = a$ and $G_x(x) = B$. We can now write down the necessary conditions of the standard result. There is one small notational difference. Since in this problem we shall have occasion to consider the multipliers as variables, we denote their particular values corresponding to the solution of the problem at hand by placing bars over the corresponding symbols.

The conditions are sufficient, too, on account of concavity. They are

$$a - \bar{\pi}B + \bar{\rho} = 0 \tag{6.11}$$

$$\bar{\pi} \geqslant 0, B\bar{x} \leqslant c \quad \text{with complementary slackness} \tag{6.12}$$

$$\bar{\rho} \geqslant 0, \quad \bar{x} \geqslant 0 \quad \text{with complementary slackness} \tag{6.13}$$

Now define $\bar{y} = c - B\bar{x}$, and using (6.11), write this, (6.12), and (6.13) in the equivalent form

$$-c + B\bar{x} + \bar{y} = 0 \tag{6.14}$$

$$\bar{x} \geqslant 0, -\bar{\pi}B \leqslant -a \quad \text{with complementary slackness} \tag{6.15}$$

$$\bar{y} \geqslant 0, \quad \bar{\pi} \geqslant 0 \quad \text{with complementary slackness} \tag{6.16}$$

Except for an interchange of rows and columns, these are exactly like (6.11)–(6.13), and are therefore necessary and sufficient conditions for the problem of choosing variables π to maximize $-\pi c$, subject to the constraints

$$-\pi B \leqslant -a, \ \pi \geqslant 0,$$

i.e. to minimize
$$\Phi(\pi) = \pi c \tag{6.17}$$

subject to

$$\pi \geqslant 0 \tag{6.18}$$

and

$$\Gamma(\pi) = \pi B \geqslant a. \tag{6.19}$$

We see at once that the new linear programming problem defined by (6.17), (6.18) and (6.19) stands in a very symmetric relation to the earlier one defined by (6.8), (6.9) and (6.10). It is customary to call the new problem the *dual* of the original one, which is then called the *primal*. The change from maximization to minimization, and the interchanges of rows and columns, and of coefficients in the objective function and the right hand sides of the constraints, are all features that should be obvious on inspection of the statements of the problems. On inspection of the respective optimization conditions, we see a more interesting interchange of the choice variables and the multipliers. The optimum choices $\bar{x}$ for the primal become the multipliers for the dual, and vice versa for $\bar{\pi}$. Also, $\bar{y}$ is the vector showing the gaps between resource availabilities c and uses $B\bar{x}$ for the primal, and $\bar{\rho}$ is the vector serving the same purpose for the constraints of the dual. We now see that $\bar{y}$ serves as the multipliers for the non-negativity constraints for the dual, and vice versa for $\bar{\rho}$.

Complementary slackness enables us to obtain another interesting relation. Consider the conditions (6.12). For any component i, we have either

$$(B\bar{x})_i = c_i \qquad \text{and therefore} \qquad \bar{\pi}_i(B\bar{x})_i = \bar{\pi}_i c_i,$$

or

$$\pi_i = 0, \qquad \text{and once again} \qquad \bar{\pi}_i(B\bar{x})_i = \bar{\pi}_i c_i (= 0).$$

Adding these over i to obtain the matrix product of the vectors, we have $\bar{\pi}B\bar{x} = \bar{\pi}c$. Similar arguments apply to (6.15), thus yielding

$$a\bar{x} = \bar{\pi}B\bar{x} = \bar{\pi}c. \tag{6.20}$$

Thus the maximum value of the primal is equal to the minimum value of the dual.

This also provides a sufficient condition for solution of linear programming problems. Thus, if we succeed in finding feasible choices $\bar{x}$ and $\bar{\pi}$ for the respective problems such that $a\bar{x} = \bar{\pi}c$, then the choices are optimum, each for its own problem. To see this for the primal, consider any x satisfying (6.9) and (6.10). Since $\bar{\pi}$ is non-negative, we can multiply each component inequality in (6.10) by the corresponding component of $\bar{\pi}$ and add to find $\bar{\pi}Bx \leqslant \bar{\pi}c$. Similarly, since $\bar{\pi}$ satisfies (6.19) and x is non-negative, we have $\bar{\pi}Bx \geqslant ax$. Then, for any feasible x, we have $ax \leqslant a\bar{x}$, which is the result. The same argument applies to the dual. This is sometimes a useful trick for finding the solutions to such problems.

This is in essence the duality theory of linear programming except for one point. We have paid no attention to the problem of existence of solutions. The problem can arise because the constraints may be mutually inconsistent, or because they may define an unbounded feasible set in a direction which makes the objective function unbounded over this set. Here, too, a duality obtains. If the primal is infeasible, then the dual is either infeasible or unbounded, and similarly the other way around. If both are feasible, then both have optimum solutions and the earlier theory is valid. I shall omit the discussion of this.

Finally, it is easy to see that if we take the dual as our starting point and go through the mechanical steps of finding *its* dual, we return to the primal. In other words, duality is 'reflexive'.

An important economic question is the interpretation to be assigned to $\bar{\rho}$. In our usual interpretation of the problem as one of production, when $\bar{x}$ yields the optimum output levels and $\bar{\pi}$ the shadow prices of the resources, a natural interpretation is available. The jth component of the left hand side in (6.19) is $\Sigma_i \pi_i b_j{}^i$. Since $b_j{}^i$ is the amount of the ith resource needed to produce a unit of good j, this is simply the shadow cost for a unit output of good j. Since a_j is the value placed by the objective function on such a unit, the constraints (6.19) amount to the requirement that at the shadow costing, no good should make a profit. This is natural, since it would have been desirable to expand production had there been such a profit, given the linearity of the problem. On the other hand, $\bar{\rho}$ is then the vector of the shadow losses in the production of various goods, and the complementary slackness conditions (6.13) say that production will not be undertaken for a good involving a shadow loss. Once again, because of linearity, the occurrence of a positive loss is a signal of the desirability of shutting down that line of production altogether. This makes economic sense, but I omit details of the argument to save space. Similarly, I must leave other aspects of linear programming, such as characterization and computation of solutions, to specialized books.

EXERCISES

6.1 State the analogues of the results of this chapter for minimization problems. Devise proofs for at least one of them.

6.2   Solve the problem of Example 4.3 using the methods of this chapter.

6.3   What conditions should be imposed on the various functions involved in the 'invisible hand' problem of Example 2.2 and Exercise 2.3 if the conditions found to be necessary for optimization there are also to be sufficient according to the results of this chapter?

FURTHER READING

For a detailed account of linear programming and its applications to economic theory and practice, the best reference is still
DORFMAN, R., SAMUELSON, P. A. and SOLOW, R. M. *Linear Programming and Economic Analysis*, McGraw-Hill, New York, 1958, especially chs. 1—7.
A more advanced treatment of optimization with inequality constraints with applications in economics can be found in
INTRILIGATOR, M. D. *Mathematical Optimization and Economic Theory*, Prentice-Hall, Englewood Cliffs, N.J., 1971, chs. 4, 5.
The classic article on the necessary conditions for optimization with no concavity requirements is
KUHN, H. W. and TUCKER, A. W. 'Nonlinear programming', in *Proceedings of the Second Berkeley Symposium on Mathematical Statistics and Probability*, ed. J. Neyman, University of California Press, Berkeley, Cal. 1950 (pp. 481—92).
For sufficient conditions in maximization of quasi-concave functions, see ARROW, K. J. and ENTHOVEN, A. C. 'Quasi-concave Programming', *Econometrica, 29*(4), October 1961, pp. 779—800.
For a mathematician's treatment of the issues of this chapter and earlier ones, and a discussion of algorithms for numerical computation of optimum choices, see
LUENBERGER, D. G. *Introduction to Linear and Nonlinear Programming*, Addison-Wesley, Reading, Mass., 1973.
A more advanced treatment can be found in
LUENBERGER, D. G. *Optimization by Vector Space Methods*, Wiley, New York, 1969, chs. 7 and 9.
For an example of an economic problem requiring a sophisticated constraint qualification, see
DIAMOND, P. A. and MIRRLEES, J. A. 'Optimal Taxation and Public Production: II Tax Rules', *American Economic Review*, LXI(3), June 1971, pp. 261—78, Section X.
An illustration of the important two-stage maximization problem discussed in this article will be considered later.

A useful reference that will help in avoiding pitfalls in optimization theory and practice is

SYDSAETER, K. 'Letter to the Editor on Some Frequently Occuring Errors in the Economic Literature concerning Problems of Maxima and Minima', *Journal of Economic Theory, 9*(4), December 1974, pp. 464—6.

7. Comparative Statics

The concept of comparative statics was introduced in Chapter 2, and some examples of it have appeared in earlier chapters. The interpretation of Lagrange multipliers as shadow prices was based on comparative static considerations, and the proof of concavity of the maximum value function $V(c)$ in Chapter 5 was also of this nature. I now turn to comparative statics for more general parametric variations. The general results are rather weak in an abstract context, but have many and varied applications. This produces a chapter with a brief text and lengthy examples.

In the notation of Chapter 3, let b be any vector of parameters. Consider the problem of maximizing $F(x, b)$ subject to $G(x, b) \leqslant 0$, and let $V(b)$ be the maximum value. Then we have the following general result —

> If F is concave and G convex, in each case jointly as a function of the choice variables and parameters, then V is concave.

The proof follows the line used so often before. Let b and b' be any two values of the parameter, and $\bar{x}$ and $\bar{x}'$ the corresponding optimum choices. Thus $G(\bar{x}, b) \leqslant 0$, $V(b) = F(\bar{x}, b)$ and $G(\bar{x}', b) \leqslant 0$, $V(b') = F(\bar{x}', b)$. Now let δ be any number such that $0 \leqslant \delta \leqslant 1$, and consider the choice $\delta\bar{x} + (1 - \delta)\bar{x}'$. Since G is convex, we have $G(\delta\bar{x} + (1 - \delta)\bar{x}', \delta b + (1 - \delta)b') \leqslant \delta G(\bar{x}, b) + (1 - \delta)G(\bar{x}', b') \leqslant 0$ so the proposed choice is feasible. Also, for it, by the concavity of F,

$$F(\delta\bar{x} + (1 - \delta)\bar{x}', \delta b + (1 - \delta)b') \geqslant \delta F(\bar{x}, b) + (1 - \delta)F(\bar{x}', b')$$

$$= \delta V(b) + (1 - \delta)V(b').$$

Then $V(\delta b + (1 - \delta)b')$ can be no less than the right hand side.

This result has its uses. As a simple example, the case of the function $V(c)$ in Chapter 5 is a special case of it. Another application concerns sufficient conditions of optimization in a dynamic context. However, it is a weak result, because the requirements of the concavity of F and convexity of G jointly in choice variables and parameters are often not fulfilled in economic problems. For example, budget constraints are not jointly convex in quantities and prices, and utility and production functions are not jointly concave in quantities and other parameters. Therefore we have to seek more specific results.

As an example, consider the case where the parameters do not affect the constraints. Then the choice that is optimum for one set of parameter values remains feasible for any other set, and this fact allows some very simple and useful value comparisons. One such result is the following —

> Let $V(b)$ denote the maximum value of $F(x, b)$ subject to $G(x) \leqslant 0$. If F is convex as a function of b alone for any fixed value of x, then V is convex.

To see this, write $V(b) = F(\bar{x}, b)$ and $V(b') = F(\bar{x}', b')$ as usual. Let $0 \leqslant \delta \leqslant 1$, and consider the weighted average parameter values, $\delta b + (1 - \delta)b'$. Suppose x^* is the optimum choice for this set. Since x^* is feasible when $\bar{x}$ or $\bar{x}'$ is chosen, we must have

$$F(x^*, b) \leqslant F(\bar{x}, b) \qquad \text{and} \qquad F(x^*, b') \leqslant F(\bar{x}', b').$$

Then, using convexity of F as a function of b, we have

$$V(\delta b + (1 - \delta)b') \leqslant F(x^*, \delta b + (1 - \delta)b')$$
$$\leqslant \delta F(x^*, b) + (1 - \delta)F(x^*, b')$$
$$\leqslant \delta F(\bar{x}, b) + (1 - \delta)F(\bar{x}', b')$$
$$= \delta V(b) + (1 - \delta)V(b')$$

One very interesting feature of this result is that no conditions had to be imposed on G. Convexity, or even quasi-convexity, in the choice variables, such an important condition in similar proofs earlier, is not required here. Of course some conditions will be necessary to ensure the existence of a solution, but given that, the fact that the feasible set remains unchanged as parameters vary is all we need.

This result has some important applications that will be considered in the examples to follow.

Another simple value comparison is possible in this case, and it enables us to deduce some properties of the optimum choice itself, again without having to impose any conditions other than those needed to ensure a solution. Consider the same problem of maximizing $F(x, b)$ subject to $G(x) \leqslant 0$, and suppose $\bar{x}'$ and $\bar{x}''$ are the optimum choices for parameter values b' and b'' respectively. Since each is feasible when the other is chosen, we must have

$$F(\bar{x}', b') \geqslant F(\bar{x}'', b') \qquad \text{and} \qquad F(\bar{x}'', b'') \geqslant F(\bar{x}', b'') \quad (7.1)$$

A similar argument is possible in the other polar case where the parameters affect the objective function but not the constraints. Consider the problem of maximizing $F(x)$ subject to $G(x, b) \leqslant 0$. In the same notation as above, suppose $\bar{x}'$ happens to be feasible when the parameters are b'', i.e. $G(\bar{x}', b'') \leqslant 0$. Since $\bar{x}''$ is the choice when $\bar{x}'$ could have been chosen, we must have $F(\bar{x}'') \geqslant F(\bar{x}')$. However, $\bar{x}'$ is chosen for parameter values b', and the only reason for not choosing $\bar{x}''$ with its higher value must be that it is then infeasible, i.e. $G(\bar{x}'', b') > 0$, Thus we have

$$\text{If } G(\bar{x}', b'') \leqslant 0, \qquad \text{then} \qquad G(\bar{x}'', b') > 0. \quad (7.2)$$

Again, applications of these results will appear in the examples that follow. These examples will illustrate the methods of comparative statics that use only the most basic concepts of optimization, namely the definitions of feasibility, optimality, concavity and convexity. This approach is very general, involves only the most elementary mathematics, and is aesthetically quite pleasing. On the other hand, few comparative static results are available at this level of generality. Most specific economic problems have more structure, i.e. the functions F and G are known or assumed to have properties besides those of concavity and convexity used in establishing the conditions for optimality. These other properties, such as additive separability, are useful in yielding further comparative static results, but the approach of this chapter is not very suitable for handling them.

The next chapter will introduce a complementary way of doing comparative statics. It begins with the $(m + n)$ equations which define the values of the n variables and the m multipliers, and differentiates these equations with respect to the parameters to find the rates of change in the variables and the multipliers. This is messy, but mechanical. The additional conditions on the various functions in special problems are often expressed in terms of their derivatives, and are therefore suitably tackled by the differentiation approach. However, this method is restrictive, not merely because some problems may involve non-differentiable functions, but also because inequality constraints may pose problems. It is not legitimate to take derivatives of both sides of an inequality to obtain another inequality. If we are to apply this method to problems with inequality constraints, we have to know in advance which of the constraints will hold as equalities, and which ones being not binding can be ignored. Further, we must be sure that the same set will hold as equalities over the entire range of parametric variation being considered, for switches from one regime to another pose their own problems for differentiation.

The two methods thus have complementary advantages and disadvantages. We should always keep both in mind, and should be ready to use the one which is best suited to the problem at hand. Correct judgement concerning this, of course, comes only with practice.

EXAMPLES

Example 7.1 This example continues the development of consumer theory using the indirect utility function and the expenditure function. The notation of Example 3.2 is retained. I shall assume the functions to be twice differentiable. Roughly speaking, this amounts to assuming that the (direct) utility function, besides being twice differentiable, has no flat portions to its indifference surfaces.

Begin with the expenditure function. The first point is that for any fixed u, $E(p, u)$ is concave as a function of p. This is because the relevant parameter affects only the criterion function, and written in the standard maximization form, $-px$ is convex (although only just) in p for each fixed x. The standard result of the text then says that $-E(p, u)$ is convex in p, i.e. $E(p, u)$ is concave in p. The economic reason pertains to substitution in consumption. For example, as one component of p increases, the worst that could happen is that it would be necessary to maintain the old consumption plan to attain the given utility level, in which case expenditure would increase linearly with the price. If it is possible to substitute against the commodity that has become more expensive, expenditure will increase less than linearly. Of course such concavity in each direction does not prove overall concavity, but provides some economic intuition for the result proved before.

For a twice-differentiable concave function, the partial derivative with respect to any argument must be non-positive. This has an implication for the derivatives of compensated demand functions. Remembering that superscripts denote the commodity number and subscripts denote partial derivatives, (3.19) gives

$$C_j^j(p, u) = E_{jj}(p, u) \leqslant 0.$$

Thus, when any price increases, the compensated amount demanded of that commodity cannot increase, i.e. the own substitution effect is non-positive. This is a well known and important theorem of consumer theory.

The same result follows without assuming differentiability from (7.1). Consider price vectors p', p'' and the corresponding compensated demands, say $\bar{x}'$ and $\bar{x}''$. Write $p'' - p' = \Delta p$, and $\bar{x}'' - \bar{x}' = \Delta \bar{x}$. Now (7.1) gives

$$-p''\bar{x}'' \geqslant -p''\bar{x}' \qquad \text{and} \qquad -p'\bar{x}' \geqslant -p'\bar{x}''.$$

On adding these inequalities and simplifying, we find

$$\Delta p \Delta \bar{x} \leqslant 0 \qquad\qquad (7.3)$$

In case only the jth component of Δp is non-zero, this becomes

$$\Delta p_j \Delta \bar{x}_j \leqslant 0,$$

which is our result.

Let us turn to the indirect utility function. This is not amenable to standard theorems. In fact V is quasi-convex in p for given m; i.e. for given m and u, the set of vectors p satisfying $V(p, m) \leqslant u$ is convex. To show this, suppose $V(p, m) \leqslant u$, $V(p', m) \leqslant u$, and $0 \leqslant \delta \leqslant 1$. We wish to show $V(\delta p + (1 - \delta)p', m) \leqslant u$. Suppose this is false, i.e. suppose there exists a feasible x^* yielding utility $U(x^*) > u$. This exceeds the utility attainable with the actual choices with prices p and p', therefore it must be the case that x^* would not be feasible in either of those situations, i.e. $px^* > m$ and $p'x^* > m$. Now each of δ and $(1 - \delta)$ is non-negative, and not both can be zero simultaneously. Therefore $(\delta px^* + (1 - \delta) p'x^*) > \delta m + (1 - \delta)m$, i.e. $(\delta p + (1 - \delta)p')x^* > m$. Thus x^* is not feasible for the weighted average price vector. This contradiction forces us to abandon our supposition that $U(x^*) > u$, thus proving the result of quasi-convexity.

This causes some problems. Consider a two-stage maximization problem in which the government, through its tax policies, can affect consumer prices. Consumers make their optimum adjustments to these policies, and then the government, in choosing the optimum policy, takes these responses into account. This can lead to a problem in which p is being chosen to maximize $V(p, m)$ subject to some constraints. But a quasi-convex function is not a very suitable maximand, particularly when we wish to establish sufficient conditions. This issue will reappear later.

One useful property can be found from (7.2) which makes no concavity assumptions. Since prices and income affect the budget constraint but not the utility function directly, we can use this equation to write, in usual notation,

if $\quad p''\bar{x}' \leqslant m''$, $\qquad$ then $\qquad p'\bar{x}'' > m'$, $\qquad$ or

if $\quad p''\bar{x}' \leqslant p''\bar{x}''$, $\qquad$ then $\qquad p'\bar{x}'' > p'\bar{x}'$, $\qquad$ (7.4)

assuming non-satiation so that for each choice the budget constraint is binding. This has an important application. It is possible to base consumer theory on properties of demand functions rather than of utility functions, and this is held to be desirable because the former are observable and the latter are not. This is called the *revealed preference* approach to consumer theory. In such a formulation, (7.4) is not a theorem, but one of the fundamental assumptions, called the Weak Axiom of Revealed Preference. It turns out, however, that the two ways of developing consumer theory are formally equivalent, once enough assumptions are made for each theory to be of any use.

Finally we relate the indirect utility function and the expenditure function, or the uncompensated and compensated demand functions. Suppose we begin with some u, and find $m = E(p, u)$. Then we assign this m as the money income, and find the utility-maximizing choice. Except for some technical problems that arise when there are some goods with zero prices, we have the expected result, i.e. $u = V(p, m)$, and the optimum choices coincide. I shall assume this to be true, i.e. that

$$u = V(p, m) \qquad \text{if and only if} \qquad m = E(p, u) \qquad (7.5)$$

and that if m and u are so related,

$$C^j(p, u) = D^j(p, m). \qquad (7.6)$$

In particular, $m = E(p, V(p, m))$, and differentiation using (3.16) and (3.17) yields

$$\lambda\mu = 1. \tag{7.7}$$

This relationship between the Lagrange multipliers of mirror-image optimization problems has obvious economic meaning.

Finally, for fixed u, differentiate (7.6) with respect to p_k, remembering that m must change according to (7.5). The chain rule gives

$$C_k^j(p, u) = D_k^j(p, m) + D_m^j(p, m) E_k(p, u).$$

Using the definition (3.19) and (7.6), this becomes

$$C_k^j(p, u) = D_k^j(p, m) + D^k(p, m) D_m^j(p, m). \tag{7.8}$$

This relation between the substitution, income and overall effects of a price change is one of the most important results of consumer theory; it is called the Slutsky-Hicks equation. It is instructive to contrast the simple derivation above with the lengthy conventional proof which relies on direct methods alone.

Readers still unfamiliar with the notation used here should recognize the result in the form

$$\left.\frac{\partial x_j}{\partial p_k}\right|_{u \text{ constant}} = \left.\frac{\partial x_j}{\partial p_k}\right|_{m \text{ constant}} + x_k \frac{\partial x_j}{\partial m}$$

Example 7.2   This example develops the elementary theory of cost-of-living indices. Consider a consumer with given tastes. Fix a utility level u which forms the standard of living chosen as the basis of comparison. For each price vector p, we can calculate the amount of expenditure necessary in order to attain this standard quite simply as $m = E(p, u)$.

Now consider two situations, the initial or base period with prices p' and the final or current period with prices p'', and suppose the corresponding expenditures are m' and m''. It is natural to say that the cost of living has gone up if more expenditure is necessary in order to attain the target utility level in the current period than in the base period, and that it has gone down if the reverse is true. We are looking for criteria to judge this in terms of observable prices and quantities.

We know that the expenditure function is concave. Assuming it to be differentiable, changes in it are overestimated by linear approximations based on tangents. Moreover, the vector of derivatives at any point is simply the vector of compensated demands there. Write $\bar{x}' = E_p(p', u)$ and $\bar{x}'' = E_p(p'', u)$; note that p being a row vector, $E_p(p, u)$ is a column vector. Then we have

$$m'' - m' = E(p'', u) - E(p', u) \leqslant (p'' - p')\bar{x}',$$

and $\qquad m' - m'' = E(p', u) - E(p'', u) \leqslant (p' - p'')\bar{x}''.$

From these, we obtain the following sufficient conditions.

If $\qquad (p'' - p')\bar{x}' < 0, \qquad$ then $\qquad m'' < m', \qquad$ and

if $\qquad (p' - p'')\bar{x}'' < 0, \qquad$ then $\qquad m' < m''.$

In the first case, the cost of living has fallen, while in the second case it has risen. A slight change enables us to write these in the standard index number form as

If $\qquad p''\bar{x}'/p'\bar{x}' < 1, \qquad$ then $\qquad m'' < m', \qquad$ and

if $\qquad p''\bar{x}''/p'\bar{x}'' > 1, \qquad$ then $\qquad m' < m''.$

The two ratios look very much like the Laspeyres (base quantity weighted) and Paasche (current quantity weighted) price indices. But we must remember that the quantities which appear here are the compensated demands at the specified utility level u, which need not have any relation to the actual demands or utilities in either period. In normal use, the standard of comparison will be the actual utility level in one of the situations, and the quantities the actual demands in that situation. This enables us to draw conclusions that depend only on observable prices and quantities. Thus, if the Laspeyres price index is less than one, the current situation has lower cost of living as judged by the base period utility standard, while if the Paasche price index is greater than one, then the base period situation has the lower cost of living as judged by the current period utility standard.

These are really rather weak statements, for they say nothing about a very wide range of other possibilities. However, there is no way out of this, and price and quantity data alone permit only a very limited set of welfare comparisons. Besides, further complications arise when we try to take account of changes in tastes, or of distributive concerns when there are many consumers.

Example 7.3   In this example we consider aspects of production theory using methods similar to those used for consumer theory in Example 7.1.

The cost function for a producer has been defined in Exercise 3.2 as the minimum cost of production given the factor prices and the target output level. It is clear that it should have the same properties as the expenditure function. Thus it should be an increasing function of all arguments, and for each fixed output level it should be homogeneous of degree zero in the factor prices, and a concave function of them. This last property is a reflection of substitution in production: with no substitution, the cost function will be linear in the factor prices, and the greater the substitution possibility, the greater its concavity. Finally, its partial derivatives are the cost-minimizing input demands, i.e. in the notation of Exercise 3.2,

$$x = C_w(w, y). \tag{7.9}$$

Remember once again that, by the convention established in Chapter 1, the argument being a row vector, the partial derivatives form a column vector.

This leaves open the determination of output, and in the example I shall outline one possibility. Suppose there are constant returns to scale, so the cost function can be written in the form $yC(w)$, where $C(w)$ is now the minimum cost of producing one unit of output. The size of the firm is indeterminate when there are constant returns to scale, and we may take this to be the industry cost function with y as the industry output, and the corresponding factor demands for the industry will be

$$x = yC_w(w). \tag{7.10}$$

Consider a competitive equilibrium. If p is the output price, each firm's profit-maximization decision will be to equate marginal cost to price. Under constant returns to scale, this is the same as equating the average cost and price, i.e.

$$p = C(w). \tag{7.11}$$

Finally, for market-clearing, we must have

$$y = D(p), \tag{7.12}$$

where D is the industry demand function. Successive substitution from these into (7.10) will define the derived factor demands as functions of the factor prices alone.

As an illustration of the use of this model, suppose we want to know how the various factor demands change as w_k changes, the remaining factor prices being unchanged. A simple differentiation gives

$$\partial x_j / \partial w_k = y C_{jk} + D'(p) C_k C_j,$$

where the arguments of the derivatives of C are omitted for brevity. It is more instructive to write this as an elasticity. After some simplification, we find

$$(w_k / x_j) \partial x_j / \partial w_k = \theta_k (\sigma_{jk} - \eta), \tag{7.13}$$

where η is the elasticity of the industry demand curve,

$$\theta_k = (w_k x_k) / (yC)$$

is the share of the factor k in total factor cost, and

$$\sigma_{jk} = (C C_{jk}) / (C_j C_k).$$

Thus the effect of a change in w_k is composed of two parts. The first is the substitution effect: as relative prices of factors change, the cost-minimizing factor proportions change. As in consumer theory, the own substitution effect is unambiguous, since C is a concave function, and therefore $C_{kk} \leqslant 0$. However, for $j \neq k$, the effect depends on whether the factors j and k are substitutes or complements. The other term gives the output effect. An increase in w_k raises the whole marginal cost schedule, thus reducing the profit-maximizing output and thereby the demand for all factors. Of course, constant returns to scale imply that this effect is equi-proportional for all factors. In general, there can even be inferior factors for which the demand rises as output falls.

The output effect in production theory should be distinguished from the income effect in consumer theory. The former arises because it is desirable to produce less as costs increase, i.e. roughly from the side of the objective function. The latter arises because it is necessary to consume less as costs increase, i.e. on account of the constraint.

Both the substitution effect and the output effect in (7.13) contain the factor θ_k. Roughly, this is because if a factor accounts for only a small fraction of cost, then a given percentage change in its price calls for only a small adjustment on either count.

The expressions σ_{jk} are called the partial elasticities of substitution for the production process, for any pair $j \neq k$. For the case $j = k$, it is better to write (7.13) differently. Since C is homogeneous of degree one, each of its partial derivatives C_k is homogeneous of degree zero, and by Euler's theorem we have

$$\sum_{j=1}^{n} w_j C_{kj} = 0.$$

This can be written equivalently as

$$\sum_{j=1}^{n} \theta_j \sigma_{jk} = 0.$$

Then (7.13) becomes

$$-(w_k/x_k)\partial x_k/\partial w_k = \theta_k \eta + \sum_{\substack{j=1 \\ j \neq k}}^{n} \theta_j \sigma_{jk} \tag{7.14}$$

This shows that the numerical value of the own price elasticity of demand for a factor is a weighted average (since the θ_k sum to one) of various elasticities. The cost share of the factor in question multiplies the elasticity of product demand, and each remaining cost share multiplies the partial elasticity of substitution between the pair of factors involved.

This is an example of a precise model for formalizing Marshall's various laws of derived demand.

Example 7.4   Consider an economy with H households and G goods. Labour is chosen as numéraire, and the prices of the goods in wage units are p_g for $g = 1, 2, \ldots G$. There is no income-leisure choice, and the labour supplies of the households are fixed at ℓ_h for

$h = 1, 2, \ldots H$. Wage income, of course, contributes to indirect utility. This function for household h is (cf. Exercise 3.3) given by

$$V^h(p, \ell_h) = \log \ell_h - \Sigma_g \alpha_{hg} \log p_g, \tag{7.15}$$

where, for each h, $\Sigma_g \alpha_{hg} = 1$. The demand for good g by household h is easily seen to be

$$x^{hg}(p, \ell_h) = \alpha_{hg} \ell_h / p_g. \tag{7.16}$$

Then the aggregate demands are

$$x^g(p, \ell_1, \ldots \ell_H) = \Sigma_h \alpha_{hg} \ell_h / p_g. \tag{7.17}$$

The production of one unit of good g needs c_g units of labour. Therefore a production plan $(x_1, x_2, \ldots x_G)$ is feasible if

$$\Sigma_g c_g x_g \leqslant \Sigma_h \ell_h. \tag{7.18}$$

The government can choose the prices of the goods, and wishes to choose a feasible production plan by doing so in order to maximize the sum of the households' utilities. Since the ℓ_h are constant, its maximand can be seen from (7.15) to be

$$F(p) = -\Sigma_g (\Sigma_h \alpha_{hg}) \log p_g. \tag{7.19}$$

Using (7.17) and (7.18), the constraint becomes

$$\Sigma_g c_g (\Sigma_h \alpha_{hg} \ell_h) / p_g \leqslant \Sigma_h \ell_h \tag{7.20}$$

The problem mentioned in the text can now be seen in an explicit context. The objective function is not concave. We know it to be quasi-convex in general; in this case it is in fact convex. The constraint function is convex, and since the relative convexity, or more precisely, the concavity of the Lagrange expression, is what really matters, we still have some hope of proving sufficiency. Fortunately, in the special case of these functions, a simple change of variables reduces the problem to standard form. Writing $q_g = 1/p_g$, and introducing the convenient abbreviations

$$A_g = \Sigma_h \alpha_{hg}, B_g = \Sigma_h \alpha_{hg} \ell_h, \ell = \Sigma_h \ell_h, \tag{7.21}$$

the problem becomes —

$$\text{maximize} \qquad \Sigma_g A_g \log q_g \qquad (7.19')$$

$$\text{subject to} \qquad \Sigma_g c_g B_g q_g \leqslant \ell. \qquad (7.20')$$

Now the maximand is concave and the constraint convex, and the Lagrangean conditions are sufficient. Introducing a multiplier π, we have

$$A_g/q_g = \pi c_g B_g. \qquad (7.22)$$

Substituting into the constraint, we see that $\pi = 0$ would not be permissible. Then the constraint must hold with equality, and

$$1/\pi = \ell/\Sigma_g A_g = \ell/\Sigma_g \Sigma_h \alpha_{hg} = \ell/\Sigma_h 1 = \ell/H.$$

Finally, using this in (7.22), we have the solution

$$p_g/c_g = HB_g/(\ell A_g). \qquad (7.23)$$

A question of major interest is the classification of goods into those for which p_g exceeds the cost of production c_g, i.e. those that are subject to a tax, and those for which $p_g < c_g$, i.e. those which are subsidized. After some tedious algebra, we find that those goods are taxed for which there is a positive covariance across h between ℓ_h and α_{hg}. Since ℓ_h are household incomes and α_{hg} their budget shares for good g, we see that good g is taxed if, on the average, it is more important in the budgets of the rich, and subsidized if on the average it is more important in the budgets of the poor. Thus we have a model to show how commodity taxation gives some redistributive leverage

EXERCISES

7.1 For the consumer demand model of Example 7.1, prove that (a) substitution effects are symmetric, i.e.

$$C_k{}^j(p, u) = C_j{}^k(p, u),$$

and (b) a Giffen good, i.e. one with a positively sloped uncompensated demand curve, must be an inferior good, i.e. one with a negative income derivative of demand, or in symbols,

if $\quad D_i^j(p, m) > 0,$ $\quad$ then $\quad D_m^j(p, m) < 0.$

Is the converse true?

7.2 $\quad$ Express the Slutsky-Hicks equation (7.8) in elasticity form.

7.3 $\quad$ Consider the production problem of Example 7.3, with $n = 2$. Let w_1 increase, but suppose that w_2, instead of remaining unchanged, adjusts to equate the supply and demand for the second factor. Show that the elasticity of derived demand for factor 1 is given by

$$[\sigma_{12}(\eta + \epsilon_2) + \theta_1\epsilon_2(\eta - \sigma_{12})] / [\eta + \epsilon_2 - \theta_1(\eta - \sigma_{12})] \quad (7.24)$$

where ϵ_2 is the elasticity of supply of factor 2, and the other symbols are as before. Can you obtain (7.14) as a special case of (7.24)?

FURTHER READING

For the various aspects of consumer and producer theory discussed in the examples, see

COOK, P. J. 'A One-line Proof of the Slutsky Equation', *American Economic Review*, LXII (2), March 1972, p. 139.

HOUTHAKKER, H. 'Revealed Preference and the Utility Function', *Economica*, N. S. *17*(2), May 1950, pp. 159–74.

FISHER, R. M. and SHELL, K. *The Economic Theory of Price Indices*, Academic Press, New York, 1972.

DIEWERT, W. E. 'A Note on the Elasticity of Derived Demand in the N-factor Case', *Economica*, N. S. *38*(2), May 1971, pp. 192–8.

HALL, R. E. 'The specification of technology with several kinds of output', *Journal of Political Economy*, *81*(4), July/August 1973, pp. 878–92.

For more on the theory and applications of commodity taxation, see DIAMOND AND MIRRLEES, op. cit., (p. 68).

For comparison, the traditional treatment of consumer and producer behaviour can be found in Malinvaud, op. cit., (p. 23), Chs. 2, 3, or in HENDERSON, J. M. and QUANDT, R. E., *Microeconomic Theory*, second edition, McGraw-Hill: New York, 1971, Chs. 2, 3.

8. Second-Order Conditions

In this chapter we shall turn to some further results in comparative statics, and their relation to second order conditions for optimization. As explained at the end of the previous chapter, this approach relies on the differentiation of first order conditions and the constraints, and is therefore confined to problems with equality constraints. However, it can also be used in problems with inequality constraints provided we confine the variations to a range where one and the same set of constraints holds with equalities, and the rest, being not binding, can be disregarded.

The general theory is quite complicated. I shall first illustrate the relationship between comparative statics and second-order conditions in a much simpler context, then derive the conditions in a very simple constrained maximization problem, and finally state the general result and give some applications of it.

Let us begin with the simplest maximization problem, with one choice variable and no constraints. For $\bar{x}$ to maximize $F(x)$, the first-order necessary condition is

$$F_x(\bar{x}) = 0. \tag{8.1}$$

Now consider the Taylor series for $F(x)$ carried beyond the first-order terms:

$$F(x) = F(\bar{x}) + F_x(\bar{x})(x - \bar{x}) + \tfrac{1}{2}F_{xx}(\bar{x})(x - \bar{x})^2 + \ldots$$

Using (8.1), we have

$$F(x) - F(\bar{x}) = \tfrac{1}{2}F_{xx}(\bar{x})(x - \bar{x})^2 + \ldots \tag{8.2}$$

For x near enough to $\bar{x}$, the quadratic term will dominate the higher order ones. Therefore, if $F_{xx}(\bar{x})$ is positive, we will be able to find an x near enough to $\bar{x}$ for which $F(x) > F(\bar{x})$. Then $\bar{x}$ will not yield a maximum of $F(x)$ in any small neighbourhood of $\bar{x}$, and hence *a fortiori* over the whole range of variation of x. The former would be classed as a local maximum and the latter as a global maximum. Thus we have found the second-order condition necessary for both types of maxima:

$$F_{xx}(\bar{x}) \leqslant 0. \tag{8.3}$$

On the other hand, if this second-order derivative is negative, the quadratic term in (8.2) will be negative, and therefore in a small enough interval around $\bar{x}$ we will have $F(x) - F(\bar{x})$ negative, irrespective of the signs of the higher order terms. Thus, given that (8.1) holds,

$$F_{xx}(\bar{x}) < 0 \qquad (8.4)$$

is a second-order condition that is sufficient for $\bar{x}$ to give a local maximum of $F(x)$. It is possible to find global sufficient conditions using second-order derivatives, but this involves some messy calculations.

Note the difference between (8.3) and (8.4): apart from the obvious difference of a weak and a strong inequality, the former applies to local and to global maxima, while the latter applies only to local maxima. Similar remarks will apply to second-order conditions in more general contexts, when I shall concentrate on the local sufficient conditions and leave the readers to state the corresponding necessary ones.

A local maximum satisfying the second-order sufficient conditions is sometimes called a regular maximum. For an irregular maximum, when $F_{xx}(\bar{x})$ is zero, we have to look at further derivatives. Even that may not work if the function F is non-analytic (i.e. if Taylor's theorem is not valid for it.) I shall not consider these problems further.

Now suppose the problem involves a parameter b. The first-order condition is $F_x(\bar{x}, b) = 0$, so differentiating this, we have

$$F_{xx}(\bar{x}, b) \, d\bar{x} + F_{xb}(\bar{x}, b) \, db = 0$$

or

$$d\bar{x}/db = -F_{xb}(\bar{x}, b)/F_{xx}(\bar{x}, b). \qquad (8.5)$$

For a regular maximum, the denominator on the right hand side is negative, and then the sign of $d\bar{x}/db$ is the same as that of $F_{xb}(\bar{x}, b)$. We see at once how the second-order condition helps us in assessing the qualitative effects of parametric changes on the optimum choice.

As a simple illustration of this, consider a profit-maximizing firm whose demand curve, and hence revenue curve, shifts. If $R(x, b)$ is the revenue curve where x is the output and b a parameter that increases for the demand curve to shift to the right (i.e. $R_b(x, b)$ is always positive) then according to (8.5) the shift will lead to a higher optimum output level if and only if $R_{xb}(\bar{x}, b)$ is positive. This requires the parametric shift to cause an increase in the marginal revenue $R_x(x, b)$ at $\bar{x}$. This is what underlies those fond paradoxes in microeconomics where an outward shift of demand leads to a fall in output, for it is easy to arrange a shift that is an increase in average revenue but a decrease in marginal revenue at the point in question.

For a problem with many choice variables but no constraints, the second-order terms in the Taylor series are

$$\frac{1}{2} \sum_{j=1}^{n} \sum_{k=1}^{n} F_{jk}(\bar{x}) (x_j - \bar{x}_j) (x_k - \bar{x}_k) = \frac{1}{2}(x - \bar{x})^T F_{xx}(\bar{x}) (x - \bar{x}) \quad (8.6)$$

where $F_{xx}(\bar{x})$ is the symmetric square matrix of the second-order partial derivatives $F_{jk}(\bar{x})$, and the superscript T indicates the transpose of a matrix. Since $(x - \bar{x})$ is a column vector, its transpose is a row vector, and thus (8.6) is a *quadratic form*. Second-order sufficient conditions will then correspond to this always having a negative sign for $x \neq \bar{x}$, i.e. the conditions will be that the quadratic form, or its associated matrix, be negative definite. The corresponding necessary condition will require it to be negative semi-definite. Now it is well known that a matrix is negative definite if a principal minor of it formed by taking any m rows and the same m columns has the sign of $(-1)^m$. Such conditions are once again useful in doing comparative statics, since the analogue of (8.5) for the case of many choice variables is

$$d\bar{x}/db = -F_{xx}(\bar{x}, b)^{-1} F_{xb}(\bar{x}, b). \quad (8.7)$$

The inverse of a negative definite matrix is also negative definite, and the information about the signs of its minors can be combined with the knowledge of F_{xb} in specific problems to obtain some results concerning the effects of changes in parameters on choice variables. Some applications of this will be considered in the examples.

Let us turn to second-order conditions for optimization problems with constraints. Once again, necessary conditions concern non-positiveness of second-order terms, and local sufficient conditions concern their negativeness. However, these now concern the Lagrangean, and need only hold for all x near $\bar{x}$ and satisfying the constraints. For this we have to consider the theory of definiteness of quadratic forms subject to (locally) linear constraints. The general theory needs some formidable mathematical machinery, and I illustrate the principles involved only in a simple geometric context and then generalize the result. Consider the case where we have two choice variables and one constraint, and where F and G in the usual notation are both increasing functions of both choice variables. There are now three possibilities, as shown in Figure 1.1 (or 4.1) and the two cases of Figure 4.2. Regard x_2 as a function of x_1 along each of the constraint curve and the level curve of the objective function through $\bar{x}$. The two have equal slopes at $\bar{x}$, and a local maximum will be assured if the former function is more concave, or less convex, than the latter, i.e. if $d^2 x_2 / dx_1^2$ along the constraint curve is (algebraically) less than that along the level curve. It remains to express these second-order derivatives in terms of the underlying functions. This is merely carrying the differentiation of implicit functions to the second order, remembering the earlier result for the first order. Thus, for the level curve, we have

$$d^2 x_2 / dx_1^2 = d(-F_1/F_2)/dx_1$$

$$= -\frac{F_2(F_{11} + F_{12}\ dx_2/dx_1) - F_1(F_{21} + F_{22}\ dx_2/dx_1)}{F_2^2}$$

$$= -(F_2^2 F_{11} - 2F_1 F_2 F_{12} + F_1^2 F_{22})/F_2^3,$$

where the argument $\bar{x}$ at which all these derivatives are to be evaluated has been suppressed for brevity. An exactly similar expression can be derived for the constraint curve. Finally, using the first-order conditions $F_1 = \pi G_1$ and $F_2 = \pi G_2$, and remembering that we are considering a case where all the F_j and G_j are positive, the second-order sufficient condition for a local maximum can be written as

$$-G_2^2(F_{11} - \pi G_{11}) + 2G_1 G_2(F_{12} - \pi G_{12}) - G_1^2(F_{22} - \pi G_{22}) > 0.$$

The corresponding necessary condition is obtained by weakening the inequality.

It is much neater to express the sufficient condition using a determinant as

$$\det \begin{vmatrix} F_{11} - \pi G_{11} & F_{12} - \pi G_{12} & -G_1 \\ F_{21} - \pi G_{21} & F_{22} - \pi G_{22} & -G_2 \\ -G_1 & -G_2 & 0 \end{vmatrix} > 0. \qquad (8.8)$$

The conditions for the general problem with n choice variables and m constraints are direct generalizations of this. In the matrix notation already established, we form the partitioned matrix

$$\begin{bmatrix} F_{xx} - \pi G_{xx} & -G_x^T \\ -G_x & 0 \end{bmatrix}$$

evaluated, of course, at $\bar{x}$. Consider its diagonal submatrices which are formed by the last j rows and columns. We can let j range from 1 to $(m + n)$, and the submatrix for this last value of j will be the whole matrix. For low values of j, the submatrices will be singular on account of the large number of zeroes in the bottom right hand corner. But the last $(n - m)$, i.e. those with j equal to $(2m + 1)$ or higher, will not necessarily be singular. Sufficient conditions for a local maximum then impose restrictions on the signs of their determinants. The signs are required to alternate, that of the first one (i.e. that formed from the last $(2m + 1)$ rows and columns) being the sign of $(-1)^{m+1}$. It is now easy to see that (8.8) is a special case of this. With $n = 2$ and $m = 1$, there is only one submatrix involved, namely the whole matrix, since $2m + 1 = n + m = 3$. The sign of its determinant is then required to be that of $(-1)^{1+1}$, i.e. positive.

Note that the successive submatrices start from the lower right hand corner, not the top left. Thus $(F_{xx} - \pi G_{xx})$ is not involved, and $(F - \pi G)$ need not be concave. For the restricted variations dx compatible with the constraints, we can have $dx^T(F_{xx} - \pi G_{xx})\,dx$ negative without such concavity, and that is all we need. Thus we see that the sufficient conditions of Chapter 6, which use such concavity, can be overly strong, although they are sometimes more convenient to use and lead to global maxima. If $\bar{x}$ is a maximizing choice but $(F_{xx} - \pi G_{xx})$ is not definite or even semi-definite there, we will have a case where $\bar{x}$ does not maximize the Lagrange expression, but merely gives some other kind of stationary point of it. This was mentioned as a possibility in Chapter 6, and we can now see how it arises.

As usual, the second-order conditions are closely related to questions of comparative statics. Consider the standard maximization problem with equality constraints, but involving parameters in the maximand and the constraints, i.e. to maximize $F(x, b)$ subject to $G(x, b) = 0$. The solution is found by solving the equations

$$F_x(\bar{x}, b) - \pi G_x(\bar{x}, b) = 0, \qquad (8.9)$$

$$G(\bar{x}, b) = 0. \qquad (8.10)$$

The optimum choice $\bar{x}$ and the multipliers π can both change in response to a change db in the parameters b. In developing the general theory, it is simpler to let all these changes occur at once, and this is done by taking the total differentials of the above equations. For the jth equation in (8.9), we have

$$\sum_{k=1}^{n} (\partial^2 F/\partial x_j \partial x_k)\, d\bar{x}_k + \sum_{r=1}^{s} (\partial^2 F/\partial x_j \partial b_r)\, db_r$$

$$- \sum_{i=1}^{m} \pi_i \left\{ \sum_{k=1}^{n} (\partial^2 G^i/\partial x_j \partial x_k)\, d\bar{x}_k + \sum_{r=1}^{s} (\partial^2 G^i/\partial x_j \partial b_r)\, db_r \right\}$$

$$- \sum_{i=1}^{m} d\pi_i \partial G^i/\partial x_j = 0.$$

This formidable expression, and a somewhat simpler one for (8.10), can both be stated in a much more compact form using vectors and matrices. In the standard notation, we find

$$\begin{bmatrix} F_{xx} - \pi G_{xx} & -G_x{}^T \\ -G_x & 0 \end{bmatrix} \begin{bmatrix} d\bar{x} \\ d\pi^T \end{bmatrix} = \begin{bmatrix} -F_{xb} + \pi G_{xb} \\ G_b \end{bmatrix} db. \qquad (8.11)$$

Of course, it is understood that all the derivatives are to be evaluated at $(\bar{x}, b)$.

It should be no surprise that the partitioned matrix on the left hand side is the same as the one involved in the second-order conditions. These conditions once again give us some information about the solutions. Their use is not easy to demonstrate in the abstract, but some applications to particular problems will be discussed in the examples.

Finally, let us examine what happens to the envelope properties of maximum value functions when we consider second-order terms. Recall the discussion in Chapter 3, where we compared two situations with different degrees of choice. With all variables free, the optimum choice is $\bar{y} = Y(b)$ and $\bar{z} = Z(b)$ as a function of the parameters, and the maximum value is

$$V(b) = F(Y(b), Z(b), b). \qquad (3.8)$$

When the set of variables y is held fixed, the optimum choice is $Z(y, b)$ and the maximum value

$$V(y, b) = F(y, Z(y, b), b). \qquad (3.9)$$

In particular, if y is held fixed at $\bar{y}$, then $Z(\bar{y}, b) = \bar{z}$ and

$$\left. \begin{array}{l} V(\bar{y}, b) = V(b) \\ V(\bar{y}, b + \mathrm{d}b) \leqslant V(b + \mathrm{d}b). \end{array} \right\} \qquad (3.12)$$

Finally, assuming differentiability, we have the first-order result

$$V_b(\bar{y}, b) = V_b(b). \qquad (3.14)$$

To illustrate second-order results, consider the case where there is only one scalar parameter b, and it affects only the objective function. Then we know from (3.2) that

$$\left. \begin{array}{l} V_b(\bar{y}, b) = F_b(\bar{y}, Z(\bar{y}, b), b) \\ V_b(b) = F_b(Y(b), Z(b), b). \end{array} \right\} \qquad (8.12)$$

Now consider the Taylor expansion of the inequality in (3.12). We have

$$V(\bar{y}, b) + V_b(\bar{y}, b)\, \mathrm{d}b + \tfrac{1}{2} V_{bb}(\bar{y}, b)\, \mathrm{d}b^2 + \frac{1}{6} V_{bbb}(\bar{y}, b)\, \mathrm{d}b^3 + \dots$$

$$\leqslant V(b) + V_b(b)\, \mathrm{d}b + \tfrac{1}{2} V_{bb}(b)\, \mathrm{d}b^2 + \frac{1}{6} V_{bbb}(b)\, \mathrm{d}b^3 + \dots$$

Using (3.12) and (3.14) and cancelling $\mathrm{d}b^2$, this becomes

$$V_{bb}(\bar{y}, b) + \frac{1}{3} V_{bbb}(\bar{y}, b)\, \mathrm{d}b + \dots \leqslant V_{bb}(b) + \frac{1}{3} V_{bbb}(b)\, \mathrm{d}b + \dots$$

If this is to hold for all db small enough, we must have

$$V_{bb}(\bar{y}, b) \leqslant V_{bb}(b). \tag{8.13}$$

This is the basic second-order envelope result. The geometric reason behind it should be clear from Figure 3.1. If there are several parameters, we can consider them one at a time to establish inequalities like (8.13) for the second-order partial derivatives with respect to each one. If we consider them all at once, we will establish the negative semi-definiteness of the matrix $[V_{bb}(\bar{y}, b) - V_{bb}(b)]$. However, this will not generally yield any useful conditions concerning particular second-order cross partial derivatives. For suitable 'regular' maxima, we can find a strict inequality like (8.13); I shall not pursue this refinement.

We can use (8.12) to express (8.13) in terms of the underlying functions. A simple application of the chain rule gives

$$F_{bz}(\bar{y}, \bar{z}, b)Z_b(\bar{y}, b) \leqslant F_{by}(\bar{y}, \bar{z}, b)Y_b(b) + F_{bz}(\bar{y}, \bar{z}, b)Z_b(b). \tag{8.14}$$

The importance of this result is that it yields a simple comparison between the responses of the actual optimum choices to parameter changes in the two situations. In particular, if we compare the situation where all the variables $x = (y, z)$ are free with the one where they are all fixed at their initial optimum levels $\bar{x}$, we can write this as

$$F_{bx}(\bar{x}, b)X_b(b) \geqslant 0. \tag{8.15}$$

These inequalities have several useful applications.

The envelope properties discussed here and in Chapter 3 can easily be generalized. The essential comparison is between one optimization problem and another with added constraints which happen to be satisfied at the optimum for the first corresponding to one value of the parameter. Clearly the maximum value attainable when there are more constraints can never exceed that when there are fewer constraints, but will just equal the latter at the particular initial point. This gives us (3.12), and the rest follows. The type of constraints which fix a subset of variables are a special case of this. Many of the results that follow from comparisons of maximization problems with differing degrees of constraints have been referred to as examples of Le Chatelier's Principle; we shall meet one such result of economic importance in the examples that follow.

EXAMPLES

Example 8.1   Consider a firm which buys a vector x of inputs, when the corresponding row vector of input prices is w, to produce output $q = Q(x)$ and sell it to obtain revenue $R(q)$. Its profits are $F(x) = R(Q(x)) - wx$. Suppose we have a regular maximum, i.e. one satisfying the second-order sufficient condition, for the choice $\bar{x}$. Now the vector of parameters b being the column vector w^T, we find on assembling components that $F_{xb} = -I$ where I is the (n-by-n) identity matrix. Then (8.7) becomes

$$d\bar{x} = F_{xx}(\bar{x}, w)^{-1} dw^T. \tag{8.16}$$

For a regular maximum, $F_{xx}(\bar{x}, w)$ and its inverse are both symmetric and negative definite. This yields two results.

First, we have

$$dw \, d\bar{x} = dw \, F_{xx}(\bar{x}, w)^{-1} dw^T < 0,$$

and in particular, for each j, $\partial \bar{x}_j / \partial w_j < 0$. Thus each factor demand curve slopes downward.

Secondly, for any j and k, $\partial \bar{x}_j / \partial w_k = \partial \bar{x}_k / \partial w_j$, i.e. the cross effects on factor demands are symmetric.

The techniques of Chapter 7 would also have led to such results on defining a profit function and examining its properties; in fact a somewhat different case of a competitive industry under constant returns to scale was considered there. One difference is that the new assumption of a regular maximum enables us to obtain strict inequalities for the own substitution effects.

Example 8.2   Consider a consumer minimizing the expenditure required to attain a target utility level. Using (8.11) it is trivial to show the symmetry of the cross-substitution effects and not difficult to show the negativity of the own ones. A new and interesting result can be found from the second-order envelope properties, particularly in the form (8.14). Take any one price, say p_1, as the parameter b. Now the maximand is $-px$, and the vector F_{bx} has a component -1 corresponding to the first good and zeroes elsewhere. This gives a particularly simple form to (8.14). Compare two problems, in both of which x_1 is among the choice variables z, but in the second of which some good, say x_2, is fixed at its optimum level corresponding to a particular value of p_1. Now consider small changes in p_1 from this value. We have

$$-\partial x_1/\partial p_1 \Big|_{x_2 \text{ free}} \geqslant -\partial x_1 \partial p_1 \Big|_{x_2 \text{ fixed}} \tag{8.17}$$

Further, comparing each problem with the trivial one where all the x_j's are fixed, using (8.15), we see that each of the above expressions is non-negative. Thus we have the result that the absolute value of the response of any compensated demand to the price of that good is greater when the remaining goods are free to vary to their new optimum levels than when one (or more) of them are held fixed. This is an example of the Le Chatelier Principle. It leads us to the presumption that when some goods are rationed, the demands for the remaining goods will become more inelastic.

Unfortunately, the general Le Chatelier Principle seems to be too elusive a concept to be stated precisely.

EXERCISES

8.1 State the second-order sufficient conditions for a 'regular' minimization problem.

8.2 Consider the compensated demands of a consumer, taking the price of good 1 to be the only relevant parameter. Derive the appropriate form of (8.11), and solve it using Cramer's Rule. Use the second-order sufficient conditions to show that the own substitution effect for good 1 is negative. How can you use the same method to obtain the same result for the remaining goods?

8.3 Illustrate the Le Chaterlier Principle using Example 8.1.

FURTHER READING

The *locus classicus* for the use of second-order conditions in deriving meaningful economic theorems is

SAMUELSON, P. A. *Foundations of Economic Analysis*, Harvard University Press, Cambridge, Mass., 1947.

For further reflections on these, including a discussion of the Le Chatelier Principle, see Samuelson's Nobel Prize Lecture, reprinted as

SAMUELSON, P. A. 'Maximum Principles in Analytical Economics', *American Economic Review*, LXII(3), June 1972, pp. 249–62.

For an extension of this principle, with applications, see

SILBERBERG, E. 'The Le Chatelier Principle as a Corollary to a Generalized Envelope Theorem', *Journal of Economic Theory*, 3(2), June 1971, pp. 146–55.

9. Optimization Over Time

In a formal sense, the general theory of optimum choice involving time requires no new principles. The reason is that we are now considering only the taking of one such decision. The variables to be chosen will pertain to different dates, but we can always stack them together in one large vector x, and the problem will remain one of maximizing a function $F(x)$ subject to some constraints $G(x) \leqslant c$. At the time when the decision is taken, the knowledge of future tastes and technology may be very imperfect. There may also be some irreducible uncertainty about events. But all this does not change the structure of the problem. We must take account of the lack of knowledge and the other uncertainties and of our attitudes to risk in setting up the functions F and G. Flexibility in the light of alternative eventualities as we now see them may become a desirable feature of plans. But once the functions are set up, the formalism takes over. Previous decisions may have to be revised as the future unfolds, experience may tell us more about the problem, and the outcome of a sequence of such decisions may look different from what we would have expected at the outset. However, these issues are outside our present limited scope, namely the methods of taking one decision that is regarded as optimum at the time it is taken, in the light of some criterion agreed upon at that time and of the possibilities as visualized then.

The reason for studying optimization involving time as a separate topic, therefore, is not that it requires any basically new theory. Rather, it is that such problems often have a special structure which enables us to say more about their solutions. The most important aspect of this special structure is the existence of stock-flow relationships among the constraints. Some of the dated variables, which I henceforth label y with the appropriate date subscript or argument, have dimensions of a stock. Others, labelled z, have the dimensions of a flow. Thinking in terms of the usual production interpretation, economic activity in one period determines the changes in stocks from that period to the next. The feasible activity levels depend both on the stocks and the flows during this period. This gives rise to constraints of the form

$$y_{t+1} - y_t \leqslant Q(y_t, z_t, t) \qquad (9.1)$$

The simplest illustration would be one where y_t is the amount of inventory in stock at time t. If a proportion δ of this stock spoils each year, then $Q(y_t, z_t, t)$ would equal the survivals from existing stocks, $(1 - \delta)y_t$, plus net new stockbuilding, z_t. I allow the date t to affect Q separately. This may happen, for example, as improvements in storage methods make δ a decreasing function of t. Another example is one where y_t is a vector of capital good stocks, which produce net outputs $\hat{Q}(y_t, t)$, and then $Q(y_t, z_t, t) = \hat{Q}(y_t, t) - z_t$, where z_t is a vector of consumption flows. Again t can affect Q separately, representing technical progress.

In addition to constraints which govern changes in stocks, there may be constraints which bind all the variables pertaining to one date, such as

$$G(y_t, z_t, t) \leqslant 0. \tag{9.2}$$

An example would be a constraint which says that consumption cannot exceed gross output. Non-negativity constraints on stocks and flows are also included in (9.2).

Another special feature that often occurs in such problems is that the criterion function separates as a sum of functions, each of which depends on the choice variables pertaining to only one date. Thus it can be written as

$$\sum_{t=0}^{T} F(y_t, z_t, t) \tag{9.3}$$

For example, a firm maximizing the discounted present value of its stream of profits would have such an objective, and time would enter the function explicitly in the form of discount factors $(1 + r)^{-t}$ where r is the rate of interest. On the other hand, (9.3) is a debatable requirement if it is imposed on the decisions of a consumer, for it implies a restriction vividly expressed as 'the marginal rate of substitution between lunch and dinner being independent of the amount of breakfast', known as the Wan-Brezski example. However, the separable form of the objective simplifies the theory a great deal, for it implies a kind of separability in the decision process as well. I shall assume this form in this elementary exposition, and refer the interested reader to a few articles that dispense with it.

Up to now I have treated time as passing in a discrete succession of periods. For most of the work, it turns out to be much more convenient to think of it as a continuous variable. There is no real theoretical reason for preferring the one or the other. Some modifications of (9.1) to (9.3) are necessary for a continuous treatment. For example, (9.3) is a sum of a finite number of terms. Treating time continuously is like dividing the total span of time from 0 to T into more and more but shorter and shorter intervals, and letting this process go to the limit. In this limit, the sum of an infinite number of terms, each of which is infinitesimally small, is exactly what is called an integral of the function over the range $[0, T]$. Readers who are unfamiliar with integration should at this point consult some elementary definitions and operations; references for this are cited in the list of Further Reading following Chapter 1. For much of the work, it will suffice to think of integrals exactly like sums, only in a different notation. Also, with continuous time, it is conventional to write $y(t)$ instead of y_t. Then the criterion can be written

$$\int_0^T F(y(t), z(t), t)\, \mathrm{d}t. \tag{9.4}$$

The stock-flow constraints will have to be modified, introducing rates of change of stocks in continuous time. Derivatives are designed to do just that. However, it is conventional to write derivatives with respect to time by means of dots over the corresponding function instead of dashes after, i.e. as $\dot{y}(t)$ instead of $y'(t)$ which would be used if t were not time. Thus

$$\dot{y}(t) \leqslant Q(y(t), z(t), t), \tag{9.5}$$

and finally, (9.2) merely needs to be rewritten as

$$G(y(t), z(t), t) \leqslant 0. \tag{9.6}$$

There is a deeper mathematical problem in treating time as a continuous variable. All of our earlier theory was developed with a finite number of choice variables. When time is being treated continuously, the choice variables $y(t)$ and $z(t)$ for all t over $0 \leqslant t \leqslant T$ amount to a continuously infinite number. We now have to use separation theorems in infinite-dimensional spaces, and there are serious difficulties in ensuring that one of the sets being separated has a non-empty interior. To give a rigorous treatment of this needs some very powerful mathematical machinery. But the result is very simple, and the theory of Chapters 4 to 6 can be applied without any noticeable change. I shall proceed to apply it, and cite further readings for interested readers.

In much of the discussion of optimization in a static context, I used a production example for constant illustration. I shall have a similar standard interpretation for the problem defined by (9.4) to (9.6), taking $y(t)$ to be stocks of capital goods and $z(t)$ to be the current activity levels, including consumption flows. Thus F can depend on both stocks and flows. Following Irving Fisher, it has become customary to emphasize that it is the consumption stream that is the real aim of economic activity, and I shall later specialize to an example where F is independent of the stocks. I shall call the value of F the utility flow, and the criterion (9.4) the utility integral.

With a finite number of constraints, we would introduce a Lagrange multiplier for each and form the Lagrange expression in the standard form. Exactly the same thing is done here, except of course that integrals replace sums. Writing the multipliers for (9.5) as $\pi(t)$ and those for (9.6) as $\rho(t)$, we have the expression

$$\int_0^T \left\{ F(y(t), z(t), t) - \pi(t)[\dot{y}(t) - Q(y(t), z(t), t)] \right.$$

$$\left. - \rho(t)G(y(t), z(t), t) \right\} dt. \tag{9.7}$$

There may be other constraints, pertaining to the end-points. For the moment, I shall assume a very simple form for these. Suppose T is fixed, and we have an initial stock vector b_0 and a target terminal stock vector b_T. Thus the added constraints are

$$y(0) \leqslant b_0, \qquad y(T) \geqslant b_T. \tag{9.8}$$

Writing these in the standard form, and using multipliers φ_0 and φ_T respectively for the two, we must add

$$-\varphi_0 [y(0) - b_0] + \varphi_T [y(T) - b_T]$$

to (9.7) to obtain the final Lagrange expression for the problem. The
first-order conditions would be found by equating to zero the partial
derivatives of this expression with respect to each choice variable. There
will of course be a continuous infinity of such conditions.

The task is made somewhat more complicated by the appearance of
$\dot{y}(t)$ in (9.7). This has an analogue in the discrete case of (9.1), where
each y_t would appear in two constraints corresponding to two adjacent
periods. The analogue also shows what can be done about the problem,
for we can rearrange the sum so that each y_t appears in only one term
in the Lagrange expression. Thus we write

$$\pi_0(y_1 - y_0) + \pi_1(y_2 - y_1) + \ldots + \pi_{T-1}(y_T - y_{T-1})$$

$$= -\pi_0 y_0 - (\pi_1 - \pi_0)y_1 - \ldots - (\pi_{T-1} - \pi_{T-2})y_{T-1} + \pi_{T-1}y_T,$$

or

$$\sum_{t=0}^{T-1} \pi_t(y_{t+1} - y_t) = \pi_{T-1}y_T - \pi_0 y_0 - \sum_{t=1}^{T-1} (\pi_t - \pi_{t-1})y_t.$$

When this is done in smaller steps in time, we have in the limit

$$\int_0^T \pi(t)\dot{y}(t)\, dt = \pi(T)y(T) - \pi(0)y(0) - \int_0^T \dot{\pi}(t)y(t)\, dt \qquad (9.9)$$

Note that $\pi_{T-1}y_T$ is replaced by $\pi(T)y(T)$, as it should be in the limit
when the length of each period is not 1, but an infinitesimal duration.

The equation (9.9) is known as the formula for integration by parts.
Note that when using it, we have to assume that the Lagrange
multipliers $\pi(t)$ are such that π regarded as a function of time is
differentiable. It is possible to do without this restriction, but the
arguments become tedious and the extension is not of much use for our
present purpose.

Suppose the optimum choice is $\bar{y}(t)$, $\bar{z}(t)$. Assuming that the
appropriate constraint qualification is met, the first order conditions
will be satisfied. Moreover, if F is a concave function, Q is a vector
concave function and G is a vector convex function, the conditions will
be sufficient.

After integration by parts, the Lagrange expression becomes

$$\int_0^T \left\{ F(y(t), z(t), t) + \dot{\pi}(t)y(t) + \pi(t)Q(y(t), z(t), t) \right.$$

$$\left. - \rho(t)G(y(t), z(t), t) \right\} dt - \pi(T)y(T) + \pi(0)y(0)$$

$$- \varphi_0 [y(0) - b_0] + \varphi_T [y(T) - b_T] \qquad (9.10)$$

When we differentiate this, we must be careful about the end-points. $y(0)$ and $y(T)$ contribute terms to the derivative both from within the integral and from outside it. The former are infinitesimal, and when they occur together with finite terms like the latter, they can be neglected. Thus the derivatives with respect to $y(0)$ and $y(T)$ produce the conditions

$$-\pi(T) + \varphi_T = 0, \qquad \pi(0) - \varphi_0 = 0$$

or

$$\pi(T) = \varphi_T, \qquad \pi(0) = \varphi_0. \qquad (9.11)$$

Using this, the complementary slackness conditions for the constraints (9.8) can be written

$$\pi(0)[b_0 - \bar{y}(0)] = 0, \qquad \pi(T)[\bar{y}(T) - b_T] = 0. \qquad (9.12)$$

Now $\pi(0)$ and $\pi(T)$ have a ready interpretation. As usual, regard b_0 and b_T as parameters of the problem, and consider the maximum utility integral as a function of them, say $V(b_0, b_T)$. Then

$$\partial V/\partial b_0 = \varphi_0 = \pi(0), \qquad \text{and} \qquad -\partial V/\partial b_T = \varphi_T = \pi(T). \quad (9.13)$$

In other words, $\pi(0)$ is the extra benefit we can secure from having another unit of initial stocks, and $\pi(T)$ is the loss we have to suffer in order to meet a terminal requirement that is more stringent by one unit. We shall see later that an extension of this is possible for interpreting all $\pi(t)$. The complementary slackness conditions have the obvious meaning.

The remaining first order conditions are found by differentiating the terms in the integral:

$$F_y(\bar{y}(t), \bar{z}(t), t) + \dot{\pi}(t) + \pi(t)Q_y(\bar{y}(t), \bar{z}(t), t) - \rho(t)G_y(\bar{y}(t), \bar{z}(t), t) = 0$$

$$(9.14)$$

$$F_z(\bar{y}(t), \bar{z}(t), t) + \pi(t)Q_z(\bar{y}(t), \bar{z}(t), t) - \rho(t)G_z(\bar{y}(t), \bar{z}(t), t) = 0 \quad (9.15)$$

However, these can also be thought of as the first-order conditions for $\bar{y}(t)$, $\bar{z}(t)$ to maximize $F(y, z, t) + \dot{\pi}(t)y + \pi(t)Q(y, z, t)$ subject to the constraints $G(y, z, t) \leqslant 0$ for each t. Let us examine what this implies. It is clear that in order to maximize the utility integral, we would not want to maximize $F(y, z, t)$ subject to $G(y, z, t)$ for each t; this would be far too short-sighted. The choice at any point in time affects the possibilities for all subsequent points through (9.5). For example, a big splurge of consumption now would leave a much smaller capital stock tomorrow, and then a lower utility flow at subsequent dates would result. This could lead to a lower utility integral, and therefore we must balance present gains against future ones at the margin in order to attain an intertemporal optimum.

We are by now used to handling gains or losses arising on account of constraints by modifying the objective function by the appropriate shadow values. In (9.15), for example, $\rho(t)G_z(\bar{y}(t), \bar{z}(t), t)$ represents the marginal cost of z considering the constraints (9.6) which apply to the variables at time t. This suggests that the other term, $\pi(t)Q_z(\bar{y}(t), \bar{z}(t), t)$, must represent a similar marginal shadow gain from constraints (9.5). This is in fact the case. We shall soon see that, in the natural sense, $\pi(t)$ is the shadow price of stocks at time t. Now an extra unit of z leads to Q_z more units of stock a unit of time later, and thus the future consequences of the choice of z are taken into account by adding the marginal shadow value term πQ_z.

Interpretation of the choice of y would be similar except for the term $\dot{\pi}(t)$. It seems tempting to think of it as the rate of accrual of extra capital gains from having another unit of stock. However, this is not a very good way of looking at the problem. It is almost never optimum to keep the inequality in (9.5) strict: an addition to stock will be desirable in all the problems we shall meet. Then there is no real *choice* of the stock levels open to us. The choice of $z(t)$ at any instant, so to speak, fixes the stocks at the next instant through (9.5) holding as an equality. This is emphasized in the technical language of the subject by calling $y(t)$ the *state* variables and $z(t)$ the *control* variables. The state variables are then built from the initial conditions and the choice of the control variables. It is therefore better to think of the stocks as passive, and to give an active role to the shadow prices: it is $\dot{\pi}(t)$ that takes on the right value to satisfy (9.15). The condition is then interpreted as saying that on the optimum time path the shadow prices change in such a way that the marginal benefits from holding an additional unit of stocks at any point in time, including the current and future shadow scarcity costs and the capital gains as well as the utility flow, are zero. In other words, producers who have to bear the shadow costs and gain the shadow profits in a decentralized economy of this kind, will be in equilibrium when holding the stocks which the past optimum flow decisions have produced. This is a natural extension of the usual role of prices in decentralization to the case of an intertemporal economy.

It is now convenient to introduce some new notation. Define a function H, called the Hamiltonian, as follows:

$$H(y, z, \pi, t) = F(y, z, t) + \pi Q(y, z, t) \tag{9.16}$$

Now (9.15) gives the first order conditions for $\bar{z}(t)$ to maximize $H(\bar{y}(t), z, \pi(t), t)$ subject to the constraints $G(\bar{y}(t), z, t) \leqslant 0$. The Lagrangean for this static maximization problem is

$$L(\bar{y}(t), z, \pi(t), \rho(t), t) = H(\bar{y}(t), z, \pi(t), t) - \rho(t)G(\bar{y}(t), z, t).$$

The maximizing choice $\bar{z}(t)$ is naturally a function of $\bar{y}(t)$, $\pi(t)$ and t; it is called the *policy function*. Substituting in the Hamiltonian, the maximum value function, or the maximized Hamiltonian, is obtained. It is written as $H^*(\bar{y}(t), \pi(t), t)$.

Now (9.14) can be written as

$$\dot{\pi}(t) = -L_y(\bar{y}(t), \bar{z}(t), \pi(t), \rho(t), t) \tag{9.17}$$

In the static maximization problem, $\bar{y}(t)$ and $\pi(t)$ are parameters affecting the criterion and the constraints. We can therefore use (3.3) to write

$$\dot{\pi}(t) = -H_y^*(\bar{y}(t), \pi(t), t). \tag{9.18}$$

Further, using the same theorem, we have $H_\pi^* = L_\pi = Q$ evaluated at the optimum. This enables us to write (9.5), now assumed to hold as an equation, in a form that is more symmetric to (9.18) as

$$\dot{\bar{y}}(t) = H_\pi^*(\bar{y}(t), \pi(t), t). \tag{9.19}$$

These two differential equations, along with the conditions for the end-points, enable us to find the functions $\bar{y}(t)$ and $\pi(t)$. Suppose we knew $\bar{y}(0)$ and $\pi(0)$. Then it would be a simple matter in principle to solve the two differential equations. There are existence theorems and analytical methods that we do not need; at the worst we can use these equations as giving approximate rates of change over small discrete intervals t and calculate approximate solutions by iteration. As a matter of fact, we do not know $\bar{y}(0)$ and $\pi(0)$, but the complementary slackness conditions (9.12) usually contain just enough information to enable us to complete the solution. We can, for example, try different values of $\bar{y}(0)$ and $\pi(0)$, and from the resulting paths, just one will yield a pair $\bar{y}(T)$ and $\pi(T)$ such that (9.12) holds. Again, in practice, more direct methods will be available, making such tedious trial-and-error techniques unnecessary. Finally, having found $\bar{y}(t)$ and $\pi(t)$, we can easily find the optimum policy path $\bar{z}(t)$.

When F and Q are concave and G is convex, the first-order conditions are also sufficient for a maximum of the utility integral. The proof is messy, and does not introduce any new concepts beyond the static case. So I shall omit it, and merely state the results so far in a collected form for reference —

> If $\bar{y}(t)$, $\bar{z}(t)$ maximize (9.4) subject to (9.5), (9.6) and (9.8), and if the appropriate constraint qualification is satisfied, then there exist non-negative functions $\pi(t)$ such that
>
> (i) $\bar{z}(t)$ maximizes $H(\bar{y}(t), z, \pi(t), t)$ subject to $G(\bar{y}(t), z, t) \leqslant 0$,
> (ii) $\bar{y}(t)$ and $\pi(t)$ satisfy the differential equations (9.18) and (9.19),
> (iii) the complementary slackness conditions (9.12) hold.
>
> If F is concave, Q is (vector) concave and G is (vector) convex, then (i)–(iii) are together also sufficient.

This result is commonly called the Maximum Principle, and was put in this framework by Pontryagin and his associates. Since it gives such prominence to the associated shadow prices, it turns out to be very useful and instructive for solving many economic problems. Some such applications will be discussed in Chapter 11.

There is one special case of the Maximum Principle that is worth a separate mention, both because it can lead to simple solutions, and because it was historically the earlier technique used for optimization involving time. This is the case where $Q(y, z, t) = z$, i.e. where the rates of change of the stocks are themselves the control variables. Suppose there are no other constraints like (9.6). Then, replacing z by $\dot{y}$ everywhere, the condition for the maximization of the Hamiltonian is

$$F_{\dot{y}}(\bar{y}(t), \dot{\bar{y}}(t), t) + \pi(t) = 0$$

Then (9.17) can be written as a differential equation involving $\bar{y}(t)$ alone:

$$\frac{\mathrm{d}}{\mathrm{d}t} [F_{\dot{y}}(\bar{y}(t), \dot{\bar{y}}(t), t)] = F_y(\bar{y}(t), \dot{\bar{y}}(t), t) \tag{9.20}$$

This is called the Euler-Lagrange equation. The total derivative on the left hand side, when evaluated using the chain rule, will produce the second order time derivative of $\bar{y}(t)$; thus the equation is a second order differential equation. It can always be solved in terms of two parameters, and then, subject to some tricky cyclic exceptions, the parameters can be adjusted to satisfy the end-point conditions.

This method is most useful when one integral of (9.20) can be found at once, thus reducing the problem to that of solving one first-order differential equation and determining a constant from initial conditions. This can be done when any one of the three arguments y, $\dot{y}$ and t is absent from F. If $\dot{y}$ does not affect F, we have $F_y(\bar{y}(t), t) = 0$. However, there is nothing of specifically intertemporal interest in the problem in this case: it reduces to separate optimizations at each instant. The other two cases are of greater interest. If y does not affect F, (9.20) integrates directly to yield

$$F_{\dot{y}}(\bar{y}(t), \dot{\bar{y}}(t), t) = \text{constant.} \tag{9.21}$$

The value of the constant is to be determined from the end-point conditions. Finally, if t does not affect F directly, we find

$$\frac{\mathrm{d}}{\mathrm{d}t} \left\{ F(\bar{y}(t), \dot{\bar{y}}(t)) - \dot{\bar{y}}(t) F_{\dot{y}}(\bar{y}(t), \dot{\bar{y}}(t)) \right\}$$

$$= F_y \dot{\bar{y}} + F_{\dot{y}} \ddot{\bar{y}} - \ddot{\bar{y}} F_{\dot{y}} - \dot{\bar{y}} \, \mathrm{d}F_{\dot{y}}/\mathrm{d}t = 0$$

using (9.20), where it is understood that the derivatives are evaluated at the optimum. Therefore

$$F(\bar{y}(t), \dot{\bar{y}}(t)) - \dot{\bar{y}}(t)F_{\dot{y}}(\bar{y}(t), \dot{\bar{y}}(t)) = \text{constant}, \qquad (9.22)$$

which is again a first-order differential equation once the constant has been determined.

This last case has a parallel that is valid in the earlier, more general, context. Suppose we have a problem in which the forces of technical progress, discounting etc. are absent, so that time does not enter explicitly as an argument in any of the functions. Then the maximized Hamiltonian H^* also does not involve t as a separate argument on its own. Then we have

$$dH^*(\bar{y}(t), \pi(t))/dt = H_y{}^*(\bar{y}(t), \pi(t))\,\dot{\bar{y}}(t) + H_\pi{}^*(\bar{y}(t), \pi(t))\dot{\pi}(t) = 0$$

using (9.18) and (9.19). Therefore the maximized Hamiltonian is constant along the optimum path. If we know the functional form of H^*, we can draw its contours in the (y, π) space, and one of these contours will be the optimum path of y and π. The end-point conditions will then help us select it. Such a contour diagram is called a *Phase Diagram*, and we shall meet an example of it in Chapter 11.

EXAMPLES

Example 9.1   This example is to illustrate how saving decisions over a lifetime can be handled using the Maximum Principle. Consider a worker with a known span of life T, over which he will earn wages at a constant rate w, and receive interest at a constant rate r on accumulated savings, or pay the same on accumulated debts. Thus, when his capital is k, his income is $(w + rk)$. If he consumes c, capital accumulation will be given by

$$\dot{k} = w + rk - c. \qquad (9.23)$$

Thus k is the state variable and c the control variable. Suppose there are no inheritances or bequests, so that the end-point conditions are

$$k(0) = 0 = k(T). \qquad (9.24)$$

Suppose there are no other constraints on the choice, and the maximand is

$$\int_0^T \log c \, e^{-\alpha t} \, dt. \tag{9.25}$$

To use the Maximum Principle we write down the Hamiltonian

$$H = \log c \, e^{\alpha t} + \pi(w + rk - c). \tag{9.26}$$

The condition for the choice of c maximizing H is

$$c^{-1} \, e^{-\alpha t} - \pi = 0, \tag{9.27}$$

and, substituting in (9.26), the maximized Hamiltonian becomes

$$H^* = -(\log \pi + \alpha t)e^{-\alpha t} + \pi(w + rk) - e^{-\alpha t}. \tag{9.28}$$

The differential equations for k and π become

$$\dot{k} = \partial H^*/\partial \pi = w + rk - \pi^{-1} \, e^{-\alpha t} \tag{9.29}$$

$$\dot{\pi} = -\partial H^*/\partial k = -r\pi \tag{9.30}$$

It is easy to solve (9.30) to obtain

$$\pi(t) = \pi_0 \, e^{-rt} \tag{9.31}$$

where π_0 is a constant to be determined. Substituting in (9.29), it is possible to integrate this by recognizing

$$d(k \, e^{-rt})/dt = (\dot{k} - rk)e^{-rt} = w \, e^{-rt} - \pi_0^{-1} \, e^{-\alpha t}.$$

Then the value of π_0 can be found using (9.24) and the solution completed. However, some economically important facts can be found without doing this. From (9.27), we find

$$c(t) = \pi_0^{-1} \, e^{(r-\alpha)t} \tag{9.32}$$

Thus, along the optimum plan, consumption grows over the worker's life if $r > \alpha$. Since consumption and wages must balance over his whole lifetime in the sense of having equal discounted present values, this must mean that $c < w$ over the earlier years and $c > w$ for the later years; i.e. the consumer saves early on and later runs down his savings.

The opposite would happen if $r < \alpha$. Of course some institutional constraints may prevent him from having negative assets by dissaving at the beginning of his life, and of course an economy could not be in equilibrium with all consumers attempting to do so, with the result that r would change. However, these are separate issues. In the special case $r = \alpha$, the wage stream would itself constitute the optimum consumption stream and there would be exactly zero saving.

This problem is even easier to solve using the Euler-Lagrange equation. Write the maximand as

$$\int_0^T \log (w + rk - \dot{k})\, e^{-\alpha t}\, dt.$$

Then eqn (9.20) becomes

$$\left\{ \frac{d}{dt}\, \frac{-e^{-\alpha t}}{w + rk - \dot{k}} \right\} = \frac{r\, e^{-\alpha t}}{w + rk - \dot{k}}.$$

In terms of c, this becomes

$$c^{-2}\dot{c}\, e^{-\alpha t} + c^{-1}\alpha\, e^{-\alpha t} = c^{-1}r\, e^{-\alpha t},$$

or

$$\dot{c}/c = r - \alpha.$$

This integrates to a form like (9.32), and it remains to determine the constant using the end-point conditions.

Example 9.2   This example has no economic content, but has the great merit that the answer is known at the outset, enabling us to follow the techniques that much better. Also, it illustrates the point that although the independent variable t in the theory has a natural interpretation as time, any other variable such as space serving the same formal role fits into the same theory.

We will find the path of minimum length between the points $(0, 0)$ and $(1, 0)$ in the plane. Choose the horizontal coordinate as t and the vertical one as y. It is clear that any path which loops or winds cannot be of minimum length, since we can simply omit a loop or an s-shape to have a shorter path. We can therefore restrict discussion to a case where y is a (single-valued) function of t. The distance between the adjacent points (t, y) and $(t + dt, y + dy)$ being $[(dt)^2 + (dy)^2]^{1/2}$, our problem is to maximize

$$\int_0^1 -(1 + \dot{y}^2)^{1/2} \, dt,$$

with $y(0) = 0 = y(1)$.

Let us begin with the Euler-Lagrange approach. Since the integrand F is independent of y, we have the integral (9.21), which reduces to $\dot{y} =$ constant, or $\bar{y}(t) = c_1 + c_2 t$ in terms of two undetermined constants. Using the end-point conditions, we find $c_1 = 0 = c_2$, or $\bar{y}(t) = 0$ for all t, which of course gives us the straight line joining the two points in question.

To use the Maximum Principle, write $\dot{y} = z$ and define the Hamiltonian

$$H = -(1 + z^2)^{1/2} + \pi z.$$

To maximize this as a function of z, we set $H_z = 0$ and solve to obtain

$$\bar{z} = \pi/(1 - \pi^2)^{1/2}, \qquad H^* = -(1 - \pi^2)^{1/2}. \tag{9.33}$$

The two differential equations are

$$\dot{\bar{y}} = \pi/(1 - \pi^2)^{1/2}, \qquad \dot{\pi} = 0. \tag{9.34}$$

Thus π is constant, a conclusion we could also have drawn by noting that since H does not involve time explicitly, H^* must be constant. Then $\dot{\bar{y}}$ is constant, and an integration and use of the end-point conditions complete the solution as before.

EXERCISES

9.1   Solve the saving problem of Example 9.1 with the maximand changed to

$$\int_0^T U(c) \, e^{-\alpha t} \, dt,$$

where $\qquad U(c) = c^{1-\epsilon}/(1 - \epsilon), \qquad \epsilon > 0.$

9.2 Solve the saving problem of Exercise 9.1 for a rentier, who has no wage income but begins his life with an inherited capital k_0 and plans to leave a bequest of k_T. How large can k_T be before a solution becomes impossible?

FURTHER READING

For more advanced discussions of the optimization methods of this chapter, see Intriligator, op. cit, (p. 68), chs. 11−14, Luenberger, *Optimization.*, op. cit., (p. 68), chs. 8, 9.

PONTRYAGIN, L. S., BOLTYANSKII, V. G., GAMKRELIDZE, R. V. and MISCHENKO, E. F. *The Mathematical Theory of Optimal Processes*, Trans. Wiley/Interscience, New York, 1962.

GEL'FAND, I. M. and FOMIN, S. V. *Calculus of Variations*, Trans. Prentice-Hall, Englewood Cliffs, N. J., 1963.

For expositions with economic applications, see

ARROW, K. J. 'Applications of Control Theory to Economic Growth', in *Mathematics of the Decision Sciences*, Part 2, American Mathematical Society, Providence, R. I., 1968, pp. 85−119.

SHELL, K. 'Applications of Pontryagin's Maximum Principle to Economics', in *Mathematical Systems Theory and Economics*, Vol. I, eds. H. W. Kuhn and G. P. Szego, Springer-Verlag, Berlin, 1969.

DORFMAN, R. 'An Economic Interpretation of Optimal Control Theory', *American Economic Review*, LIX(5), December 1969, pp. 817−31.

We shall not need any properties of differential equations beyond those developed in the course of the exposition, but interested readers can consult

BIRKHOFF, G. and ROTA, G-C., *Ordinary Differential Equations*, Second Edition, Blaisdell, Waltham, Mass., 1969, chs. 1, 2, 4 and 5.

For a discussion of the economics of decentralization in intertemporal problems, see Koopmans, op. cit. (p. 23), pp. 105−26, Malinvaud, op. cit. (p. 23), ch. 10, and Heal, op. cit. (p. 23), ch. 12.

10. Dynamic Programming

In the previous chapter we saw how the value $\pi(0)$ could be interpreted as the vector of shadow prices of the initial stocks. It was also stated that all the $\pi(t)$ could be interpreted as shadow prices of the stocks at t. The simplest way to see this is to use that earlier result, and this is done by setting up a problem in which $\pi(t)$ become the shadow prices of initial stocks. For this, we must take t to be the starting point of the optimization problem. Consider any particular value t' of t, and consider the problem of maximizing an integral exactly like (9.4), but extending over the smaller interval $[t', T]$, subject to the constraints (9.5) and (9.6) over the same interval. Leave the terminal conditions $y(T) \geqslant b_T$ unchanged, and allow a more general initial condition $y(t') \leqslant y'$, where y' is some parametric vector. The maximum value for this problem will depend, among other parameters, on t' and y'; write it as $V(t', y')$.

Now suppose the problem of Chapter 9 for the whole interval $[0, T]$ has been solved, and the optimum paths $\bar{y}(t)$, $\bar{z}(t)$ and $\pi(t)$ obtained. Now set $y' = \bar{y}(t')$. It is easy to see that the same paths truncated and considered only over the interval $[t', T]$ satisfy all the first order conditions of the Maximum Principle over this interval. Subject to the concavity assumptions, they are sufficient to ensure the optimality of the truncated paths.

Similarly, we could have chosen a terminal time t'' short of T, and imposed a terminal stock requirement $y'' = \bar{y}(t'')$, and the portion of the optimum for the full problem would remain optimum for the subproblem. In other words, if we truncate an optimum path to any subinterval, we have an optimum path for the truncated optimization problem over the same subinterval with the end-point conditions defined by the values of the state variables at the points chosen for truncation. This is a consequence of the special structure of separability of the maximand and the constraints.

Now we can use our earlier result and interpret $\pi(t') = V_y(t', \bar{y}(t'))$ as the shadow prices of initial stocks in the subproblem. Each component of $\pi(t')$ thus shows the addition to the utility integral over $[t', T]$ that would result from a unit increase in the initial stocks $y' = \bar{y}(t')$. This is not yet the same as the addition to the utility integral over $[0, T]$ that would result from an increase in the stocks at t', for in this latter case we could anticipate such an increase and adjust the utility flows over $[0, t']$ suitably. But an envelope theorem argument again saves us from having to do this recalculation explicitly. Treating the addition to stocks at t' as parametric, the fact that the utility flows were arranged to be optimum in the initial setting enables us to say that to the first order a readjustment would make no difference. Since the particular value t' could have been chosen arbitrarily, this completes the interpretation of $\pi(t)$ as the shadow prices of the stocks at time t, for every t in $[0, T]$.

The function V has other uses besides helping to establish this interpretation. In fact we can develop the whole theory of intertemporal optimization and obtain methods of solution based on this function. This method is called Dynamic Programming. I shall derive the basic results of it using the Maximum Principle already established, but the two approaches are equivalent, and it is possible to develop the argument the other way.

As usual, $\bar{y}(t)$, $\bar{z}(t)$ and $\pi(t)$ will denote the optimum paths of the relevant variables over the whole interval $[0, T]$. Fix any t within this range, and consider $V(t, \bar{y}(t))$. Take a very short time interval dt and write, to first order,

$$\bar{y}(t + dt) = \bar{y}(t) + Q(\bar{y}(t), \bar{z}(t), t)\, dt.$$

Over the interval from t to $(t + dt)$, the contribution to the utility integral is $F(\bar{y}(t), \bar{z}(t), t)\, dt$. Thereafter, the path continues as the truncation of the full optimum to the interval $[t + dt, T]$, and hence is the optimum over this interval when the initial stocks are $\bar{y}(t + dt)$. Thus the remaining utility integral equals $V(t + dt, \bar{y}(t + dt))$. Hence

$$V(t, \bar{y}(t)) = F(\bar{y}(t), \bar{z}(t), t)\, dt + V(t + dt, \bar{y}(t + dt)), \qquad (10.1)$$

to first order. Now consider any value of z satisfying $G(\bar{y}(t), z, t) \leqslant 0$. If this had been chosen at t, we would have, to first order,

$$y(t + dt) = \bar{y}(t) + Q(\bar{y}(t), z, t)\, dt.$$

With this, we would have a contribution $F(\bar{y}(t), z, t)\, dt$ over this interval, and at most $V(t + dt, y(t + dt))$ thereafter to the utility integral. Further, the sum of these two can never exceed $V(t, \bar{y}(t))$, for if it did, we would have found a feasible policy that performed better than the one we began with as the optimum. Thus we must have

$$V(t, \bar{y}(t)) \geqslant F(\bar{y}(t), z, t) \, dt + V(t + dt, y(t + dt)). \quad (10.2)$$

Combining (10.1) and (10.2), we conclude that

$$V(t, \bar{y}(t)) = \max_z \{F(\bar{y}(t), z, t) \, dt + V(t + dt, y(t + dt))\} \quad (10.3)$$

the maximum being subject to the constraints $G(\bar{y}(t), z, t) \leqslant 0$.

We can simplify (10.3) by taking a linear approximation to the value of V on the right hand side. We have, to first order,

$$V(t + dt, \bar{y}(t) + Q(\bar{y}(t), z, t) \, dt)$$
$$= V(t, \bar{y}(t)) + V_t(t, \bar{y}(t)) \, dt + V_y(t, \bar{y}(t))Q(\bar{y}(t), z, t) \, dt$$

On substituting this in (10.3), we find that $V(t, \bar{y}(t))$ cancels. Then we can divide by dt, and finally note that the term $V_t(t, \bar{y}(t))$ does not involve z and can therefore be moved outside the maximization operator. This leaves us with

$$0 = V_t(t, \bar{y}(t)) + \max_z \{F(\bar{y}(t), z, t) + V_y(t, \bar{y}(t))Q(\bar{y}(t), z, t)\} \quad (10.4)$$

Notice that the maximand looks exactly like the Hamiltonian of (9.16), except that $V_y(t, \bar{y}(t))$ replaces $\pi(t)$. And there could not be a more natural substitution, since our shadow price interpretation of $\pi(t)$ shows that these two are equal. The maximization is subject to the same constraints as before, and this produces the maximum value Hamiltonian H^*. We can therefore write

$$V_t(t, \bar{y}(t)) + H^*(\bar{y}(t), V_y(t, \bar{y}(t)), t) = 0 \quad (10.5)$$

This is the fundamental equation of Dynamic Programming.

This equation also provides us an alternative method for solving the intertemporal optimization problem. We start, of course, without knowledge of the function V. But a purely static optimization gives us the functional form of H^*. Now we know that V must satisfy the partial differential equation $V_t + H^*(y, V_y, t) = 0$. We can solve this in terms of various undetermined constants, and then use the end-point conditions to determine the values of these constants. This is not easy in practice, since the functional form of H^* will in general be quite complex, and solution of partial differential equations is not a trivial

matter in any case. For these reasons, analytic solutions can be found
only in some very simple cases, and numerical solution depends on good
computing facilities. Consequently Dynamic Programming has
proved somewhat less useful than the Maximum Principle in solving
economic problems, especially in economic theory where closed form
solutions are often sought. It does, however, produce some shadow
price results very quickly. Also, it becomes more useful in problems
involving uncertainty.

The discussion so far has kept the terminal time T and the associated
target stock requirement b_T fixed, allowing the initial time and stock to
vary. The reverse is also possible, and leads to an equation very much
like (10.5). Write $W(t, y)$ as the maximum utility integral over $[0, t]$
with a fixed stock endowment b_0 at 0 and the requirement y at t. Now
the shadow price interpretation becomes $\pi(t) = -W_y(t, \bar{y}(t))$. The
change of sign is natural since having to meet a larger requirement
reduces the possible utility flows. Also, we must now split $W(t, \bar{y}(t))$
into a utility flow over $[t - dt, t]$ and $W(t - dt, \bar{y}(t - dt))$. This causes
another change of sign. The result is

$$W_t(t, \bar{y}(t)) - H^*(\bar{y}(t), -W_y(t, \bar{y}(t)), t) = 0. \qquad (10.6)$$

This alternative approach is suitable for extending our discussion of
optimization over time to allow more general end-point conditions. Our
choice of the initial condition fixing the time and stock availability is a
very natural one, but forms of the terminal conditions other than a
fixed date and stock requirement are conceivable. Thus we may have a
fixed target for stocks, and may wish to attain it in the shortest possible
time. Now T itself is the choice variable, and the maximand can be
written as $\int_0^T (-1)\, dt$. Again, there may be some flexibility both in the
terminal time and stocks, with a longer time being allowed in return for
meeting a more stringent requirement. A general form of such
constraints can be given by

$$J(T, y(T)) \leq 0. \qquad (10.7)$$

We can solve such a problem in two stages. First we fix T and $y(T)$, and
solve the earlier problem with the corresponding terminal conditions.
This produces a value $W(T, y(T))$ for the criterion. Among all pairs
$(T, y(T))$ satisfying (10.7), we then have to choose the one which
maximizes this. The second stage is a static problem, for which we have

a Lagrange multiplier ξ such that the optimum choice $(\bar{T}, y(\bar{T}))$ satisfies the first order conditions

$$W_t(\bar{T}, y(\bar{T})) = \xi J_t(\bar{T}, y(\bar{T})), \quad W_y(\bar{T}, y(\bar{T})) = \xi J_y(\bar{T}, y(\bar{T})).$$

Using (10.6) and the shadow price interpretation, this becomes

$$\left. \begin{array}{c} -\pi(\bar{T}) = \xi J_y(\bar{T}, y(\bar{T})) \\ H^*(\bar{y}(\bar{T}), \pi(\bar{T}), \bar{T}) = \xi J_t(\bar{T}, y(\bar{T})). \end{array} \right\} \quad (10.8)$$

So long as ξ is positive, this says that the vector $(H^*, -\pi)$ should be parallel to the vector (J_t, J_y) when both are evaluated at the optimum. Since the latter is perpendicular to the surface defined by $J(t, y) = 0$, the former should also be perpendicular to this surface. For this reason, the conditions (10.8) are called the *transversality conditions*. As an example, consider the minimum-time problem mentioned before. There T does not enter explicitly in the terminal conditions defined by (10.7). Thus J_t is identically zero, and the transversality conditions reduce to $H^* = 0$ at the optimum. On the other hand, if T is fixed but the terminal stock is totally unconstrained, we have J_y identically zero and the transversality conditions yield $\pi(T) = 0$.

Dynamic Programming, the Maximum Principle, and in special cases the Euler-Lagrange equation provide us three alternative and equivalent means of solving optimization problems involving time. They have different advantages and disadvantages, and are all useful in some economic applications. Incidentally, physicists have long been formulating laws of motion in a way that is formally very similar to economic optimization over time. All three of these approaches originate in physics. For example, physicists know (9.18) and (9.19) as the Hamiltonian canonical equation of motion, and (10.5) as the Hamilton-Jacobi equation. The variables y and π usually have the physical interpretations of position and momentum respectively. It is well worth the effort to read and compare the treatment of these problems in physics and in economics.

There is one more extension of the basic intertemporal optimization problem that is important in many economic problems. Often there is no natural way to specify a terminal date for decisions. In fact, we can rarely fix a date in advance and claim that considerations beyond it can be totally disregarded. This may be a minor problem for an individual, but becomes more and more important as we consider wider and wider contexts of decision-making. It may seem that the terminal stock requirement in a finite-time problem is designed to take account of the indefinite future, for the role of such stocks is implicitly to provide utility flows beyond the terminal date. But it is precisely this that makes such a procedure very inadequate. The level of the target stock requirement will have to be fixed in a purely arbitrary way without an explicit analysis of the subsequent utility flows. But such an explicit account means solving a problem very much like the original one but with a longer time horizon. Of course, there is no logical stopping point to this argument. This forces us to allow an infinite time horizon.

We run into some technical problems when we consider decisions over an infinite time horizon. First, there is the possibility that the utility integrals may not always be finite. If two or more feasible plans each yield an infinite utility integral, we cannot directly say which is preferable. Thus our old definition of an optimum as a plan providing the highest utility integral becomes useless. Comparisons between two infinite integrals can only be made partially and in a roundabout way. The simplest method is to compare the integrals taken over the same finite horizon and then take limits as the common horizon goes to infinity. Thus, for two plans $(y(t), z(t))$ and $(y'(t), z'(t))$ over the infinite future, and for each T, we calculate

$$\int_0^T F(y(t), z(t), t) \, dt - \int_0^T F(y'(t), z'(t), t) \, dt, \qquad (10.9)$$

and take the limit of this difference as T goes to infinity. If the limit is non-negative, we say that the plan $(y(t), z(t))$ *overtakes* $(y'(t), z'(t))$. If a feasible plan overtakes all other feasible plans, it is an optimum.

In a situation with convergent utility integrals this must reduce to the old definition of an optimum, for then the limit of the difference is the difference between the limits. Thus the overtaking criterion is no less general. But the positive advantage from using it is only partial. We cannot compare all pairs of infinite utility integrals using it, for it is quite possible to have cases where the difference (10.9) goes on oscillating repeatedly between positive and negative values. Nothing can be said about the relative merits of such plans using the overtaking concept. Still, there are cases where the approach helps, and it, or some subtle variations of it, have become common.

There is another reason why an optimum may fail to exist. Discontinuities provide a reason why no path can overtake all others; thus we may be able to find a sequence of paths each of which overtakes its predecessor, but such that the limiting path of the sequence, instead of being optimum, is very undesirable. A typical example of this situation is the following. Let consumption and investment be perfect substitutes in production. Suppose the marginal productivity of investment is constant and equal to b, and suppose we wish to maximize the integral of consumption discounted back to time 0 at a discount rate r, with $r < b$. If we divert a unit of output from consumption to investment at time 0 and let it compound up to time T, the added amount of consumption available will be $\exp(bT)$, and its discounted present value will be $\exp(b - r)T$. This exceeds the opportunity cost of the investment, namely 1. Thus any investment increases the utility integral, and we can find a sequence of more investment allowed to mature longer, yielding progressively higher values of the criterion. The limit of such a sequence would be all investment and no consumption ever, which is the worst of all policies.

There are other, more subtle, kinds of discontinuities, but it seems best to leave those to more advanced expositions.

We can now consider the conditions for an optimum. It is clear that the principle that a portion of an optimum path must remain optimum for the appropriately formulated subproblem is still valid. For any finite subproblem, moreover, we can use our earlier definition of optimality and the corresponding conditions such as the Maximum Principle or Dynamic Programming. Using the former, for example, we see that the control variables should maximize the Hamiltonian at each instant in time, and the state variables and the shadow prices should satisfy the differential equations (9.18) and (9.19). But the transversality conditions present a problem. Over any finite interval with the appropriate stock constraints at its end-points we will have the standard complementary slackness conditions, and thus the condition corresponding to the initial point $t = 0$ will remain valid. However, there is no terminal stock requirement at $t = \infty$. We might think that something could be gained from a constraint that stocks must be non-negative at each point in time. Thus we might set up a finite time horizon problem where the target stock has to be at least zero, i.e. $b_T = 0$. Then the transversality condition is $\pi(T)\bar{y}(T) = 0$. Hence we might conjecture that for the infinite horizon problem the terminal condition would be

$$\lim_{t \to \infty} \pi(t)\bar{y}(t) = 0. \tag{10.10}$$

However, in general this condition is not necessary. It is only for problems in a somewhat limited range that it can be shown to be a necessary condition. It has a much more important and generally valid role as a sufficient condition when taken together with the standard concavity requirements and the other two conditons of Hamiltonian-maximization and the differential equations. The proof is a simple application of all the techniques used so far and of the overtaking definition, but it has little economic interest, and I relegate it to an example following this chapter.

EXAMPLES

Example 10.1 As an illustration of Dynamic Programming and transversality conditions, consider the minimum-distance problem of Example 9.2, modified so that the terminal point is allowed to lie anywhere on the line $t = 1$. As explained in the text, this can be done in two parts. First we find the path of minimum length from $(0, 0)$ to some point $(1, y(1))$. This leads to equations (9.33) and (9.34) as before, and it remains to find the value of π from the equation

$$\pi/(1 - \pi^2)^{1/2} = y(1). \tag{10.11}$$

Finally, to determine the choice of $y(1)$, we employ the transversality condition. From (10.8) and the fact that the terminal constraint is independent of y, the condition becomes $\pi(1) = 0$, and then from (10.11) we have $y(1) = 0$. Thus we choose the terminal point and the path in such a way that at this point the path is perpendicular to the curve on which the point is constrained to lie.

The problem can in fact be tackled from first principles, combining the two stages into one. Define $V(t, y)$ as the shortest distance from the point (t, y) to the line $t = 1$. Exactly the same argument as in the text establishes (10.5) for this case, where H^* is defined as in (9.33). Thus we have

$$V_t - (1 - V_y{}^2)^{\frac{1}{2}} = 0. \qquad (10.12)$$

This is subject to the obvious boundary condition that $V(1, y) = 0$ for all y.

While the general methods for solving partial differential equations are quite hard, it is easy to verify in this instance that $V(t, y) = 1 - t$ is the desired solution, given the appropriate choice of the sign of the square root. Of course this value of the distance is attained by the perpendicular from the point (t, y) on to the line $t = 1$.

Example 10.2   This example sketches the proof of sufficiency in the infinite context, in the sense of the overtaking criterion. Suppose that for each t, regarding (y, z) as one vector argument, F is a concave function, Q a vector concave function, and G a vector convex function. Suppose further that there exist non-negative multipliers $\pi(t)$ and $\rho(t)$ such that, omitting time arguments for brevity where no ambiguity is possible, we have

$$F_y(\bar{y}, \bar{z}, t) + \dot{\pi} + \pi Q_y(\bar{y}, \bar{z}, t) - \rho G_y(\bar{y}, \bar{z}, t) = 0 \qquad (9.14)$$

$$F_z(\bar{y}, \bar{z}, t) + \pi Q_z(\bar{y}, \bar{z}, t) - \rho G_z(\bar{y}, \bar{z}, t) = 0 \qquad (9.15)$$

for each t, and

$$\lim_{T \to \infty} \pi(T)\bar{y}(T) = 0. \qquad (10.10)$$

Then $\bar{y}(t)$, $\bar{z}(t)$ yield the overtaking optimum of

$$\int_0^\infty F(y(t), z(t), t)\, dt$$

subject to a given $y(0)$, and

$$\dot{y} = Q(y, z, t) \qquad (9.5)$$

$$G(y, z, t) \leqslant 0. \qquad (9.6)$$

Note that we have subsumed the differential equations (9.18) and (9.19) in (9.5) and (9.14) here

Further, (9.6) subsumes non-negativity conditions.

To prove the result, begin by observing that for each t, $F + \dot{\pi}y + \pi Q$ is a concave function of (y, z). Thus, from the sufficiency result of Chapter 6, we have in (9.14) and (9.15) the necessary and sufficient conditions for $(\bar{y}, \bar{z})$ to maximize this subject to the constraints (9.6) involving the convex function G. Therefore, for any (y, z) satisfying these, we have

$$F(\bar{y}, \bar{z}, t) + \dot{\pi}\bar{y} + \pi Q(\bar{y}, \bar{z}, t) \geqslant F(y, z, t) + \dot{\pi}y + \pi Q(y, z, t)$$

Now suppose we have $(y(t), z(t))$ satisfying (9.5) as well. Then

$$F(\bar{y}, \bar{z}, t) + \dot{\pi}\bar{y} + \pi\dot{\bar{y}} \geqslant F(y, z, t) + \dot{\pi}y + \pi\dot{y}.$$

This holds for each t, and can be integrated from 0 to T. Since we have $\mathrm{d}(\pi y)/\mathrm{d}t = \dot{\pi}y + \pi\dot{y}$, etc, this yields

$$\int_0^T F(\bar{y}, \bar{z}, t)\,\mathrm{d}t + \pi(T)\bar{y}(T) - \pi(0)\bar{y}(0) \geqslant$$

$$\int_0^T F(y, z, t)\,\mathrm{d}t + \pi(T)y(T) - \pi(0)y(0).$$

Using the fact that $\bar{y}(0) = y(0)$, and the non-negativity of π and y, and finally the transversality condition (10.10), we have on taking limits

$$\lim_{T \to \infty} \int_0^T F(\bar{y}(t), \bar{z}(t), t)\,\mathrm{d}t - \int_0^T F(y(t), z(t), t)\,\mathrm{d}t \geqslant 0.$$

Since $(y(t), z(t))$ could be any feasible policy, this proves the result.

EXERCISES

10.1   Consider the rentier of Exercise 9.2, now planning over an infinite horizon, starting with capital stock k at time t. Show that

$$V(t, k) = \left[\frac{\alpha - r(1 - \epsilon)}{\epsilon}\right]^{-\epsilon} \frac{k^{1-\epsilon}}{1 - \epsilon} e^{-\alpha t}$$

solves the fundamental equation of Dynamic Programming for his optimization problem. Deduce that his optimum policy is one of saving a constant fraction of his income at all times. Under what restrictions on the parameters will the transversality condition be satisfied?

10.2   Solve the problem of Example 10.1 by defining $W(t, y)$ as the shortest distance from $(0, 0)$ to (t, y), and using (10.6).

FURTHER READING

For a more advanced treatment of Dynamic Programming and its relation to other methods of intertemporal optimization, see Intriligator, op. cit., ch. 13, and mathematically more rigorous treatments in

DREYFUS, S. *Dynamic Programming and the Calculus of Variations,* Academic Press, New York, 1966.

BELLMAN, R. and KALABA, R. *Dynamic Programming and Modern Control Theory,* Academic Press, New York, 1965.

For a classic treatment, see

BELLMAN, R. *Dynamic Programming,* Princeton University Press, Princeton, N. J., 1957.

For a discussion of problems that arise in infinite-horizon planning models see Heal, op. cit. (p. 23), chs. 11, 13. A model where the transversality condition (10.10) is also necessary is developed by

WEITZMAN, M. L. 'Duality Theory for Infinite Horizon Convex Models', *Management Science, 19*(7), March 1973, pp. 783–9.

Intertemporal optimization methods are applied to physics in most modern books on mechanics; my favourite is

SYNGE, J. L. and GRIFFITHS, B. A. *Principles of Mechanics,* Third Edition, McGraw-Hill, New York, 1959.

11. Some Applications

This chapter is devoted to an exposition of two examples of economic models using the techniques developed in the previous chapters, and a statement of some others as exercises and further readings.

Example 11.1   This example considers the problem of optimum saving in a manner similar to that of Exercise 9.1, but from the point of view of society as a whole. This produces two new features. First, there is no logical terminal date to the plan. Secondly, the rate of return to saving cannot be taken to be fixed by exogenous market forces, but will depend on accumulated capital.

The simplest case of this occurs in a one-good model, where physical capital stock k consists of accumulated savings. Output flow is then $F(k)$, where F is an increasing and strictly concave function, with $F(0) = 0$ and $F'(0) = \infty$. Capital depreciates at a rate δ. If consumption flows take place at a rate c, then capital accumulation is given by

$$\dot{k} = F(k) - \delta k - c. \tag{11.1}$$

The initial capital stock $k(0)$ is given. Suppose there are no other constraints, and suppose the aim of the plan is to maximize

$$\int_0^\infty U(c) \, e^{-\alpha t} \, dt \tag{11.2}$$

in the overtaking sense, where U is an increasing strictly concave function. Detailed comments on these assumptions can be found in economic texts cited later.

This allows a straightforward application of the Maximum Principle. We define the multipliers over time, $\pi(t)$, and the Hamiltonian

$$H = U(c) \, e^{-\alpha t} + \pi(t) \, (F(k) - \delta k - c). \tag{11.3}$$

All the concavity conditions are satisfied, so the first-order conditions are necessary and, along with transversality, sufficient. Henceforth I shall consider all variables only at their optimum values, and drop the bars distinguishing them as such for sake of brevity.

Maximization of the Hamiltonian yields, for each t,

$$U'(c)\, e^{-\alpha t} = \pi. \tag{11.4}$$

The differential equation satisfied by π is

$$\dot\pi = -\pi(F'(k) - \delta). \tag{11.5}$$

We could use (11.4) to substitute for c in (11.1) to obtain a pair of differential equations for k and π. However, in this case it is easier to do the reverse, i.e. substitute for π in (11.5) to give a pair of differential equations in k and c. We have

$$U''(c)\dot c\, e^{-\alpha t} - U'(c)\alpha\, e^{-\alpha t} = -U'(c)\, e^{-\alpha t}(F'(k) - \delta).$$

To simplify this, define

$$\epsilon(c) = -cU''(c)/U'(c). \tag{11.6}$$

This enables us to write

$$\dot c/c = [F'(k) - (\alpha + \delta)]/\epsilon(c). \tag{11.7}$$

Observe that Example 9.1 and Exercise 9.1 had a formally identical structure, with $F'(k)$ constant and equal to r, and $\epsilon(c)$ constant (equal to 1 in Example 9.1 and to ϵ in Exercise 9.1). This is why they share the convenient property of a constant rate of growth of consumption along the optimum path.

The pair of equations (11.1) and (11.7) has the convenient property that t does not enter explicitly on the right hand side. Thus, if we are given any pair (k, c), we shall be able to find the rates of change $(\dot k, \dot c)$ from these equations. In the (k, c) plane, these velocities can be shown by a small vector arrow attached to the point (k, c). If we do this for all points, we can join successive arrows together to find all paths $(k(t), c(t))$ which satisfy the two differential equations, both necessary conditions for optimality. No two such paths can cross, since the direction of motion is unique given the starting point. If, from among all such paths, for a given $k(0)$ we are able to find a $c(0)$ such that the path starting there satisfies the transversality condition, we will have found an optimum.

Figure 11.1 shows this diagram. The easiest way to understand it is to think of the plane as being split into regions where the directions of change of the variables are the same. Since each of the two variables k and c can increase or decrease, we have four possible combinations, and indeed in this case there are four regions. From (11.1), we see that k increases if $c < F(k) - \delta k$, which is the region below the curve

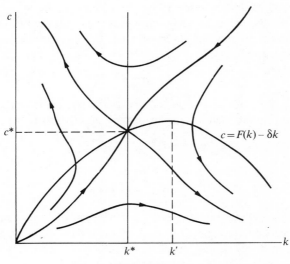

FIG. 11.1

$c = F(k) - \delta k$ in the diagram. This curve has its peak where $F(k) - \delta k$ is maximum, i.e. for $k = k'$ defined by $F'(k') = \delta$. Turning to (11.7), we see that c increases if $F'(k) > \alpha + \delta$, since $\epsilon(c)$ is positive when $U''(c)$ is negative. Now define the vertical line $k = k^*$ by $F'(k^*) = \alpha + \delta$, so that c increases to its left and decreases to its right. Since F is a concave function, we have $k^* < k'$ when $\alpha > 0$. It is then easy to verify that all possible paths satisfying the two differential equations fall into one of the patterns shown by the arrows.

Writing $c^* = F(k^*) - \delta k^*$, we see that there are exactly two of these paths which converge to (k^*, c^*), and that for each $k(0)$, there is exactly one $c(0)$ providing the initial point on one of these two. Suppose we make this choice. Then in the limit, $k(t)$ goes to k^*, and $\pi(t)$ goes to $U'(c^*)\, e^{-\alpha t}$. Provided α is positive, the transversality condition is satisfied and the choice is optimum.

Had we used the variables k and π, the diagram would have been a proper phase diagram in the sense explained in Chapter 9. In fact economists often call a diagram involving solutions to a pair of differential equations that do not involve time explicitly a phase diagram even when the two variables do not stand in the relation of a quantity and its shadow price. However, in this case the equations in k and π do involve time explicitly through the discount factor. In the case of exponential discounting, a very simple change of variables eliminates this. Define $\psi(t)$ by

$$\pi(t) = \psi(t)\, e^{-\alpha t}.$$

Comparing this with (11.4), we see that $\psi(t)$ is the undiscounted shadow price of $k(t)$. Now a mechanical differentiation shows

$$\dot{\psi} = -\psi\,[F'(k) - (\alpha + \delta)] \tag{11.8}$$

and

$$\dot{k} = F(k) - \delta k - V(\psi), \tag{11.9}$$

where V, the inverse function to U', is a decreasing function since U' is. It is now easy to draw the phase diagram in terms of k and ψ. Since c and ψ are inversely related, this looks exactly like Figure 11.1 but reflected upside down. This is left as a simple exercise.

The study of optimum saving policy forms an obviously important part of growth theory. Only the very briefest sketch of the simplest such model can be accommodated here, but the interested reader has a very extensive literature available for pursuing further developemnts. A sample of such work is listed at the end of the chapter.

Example 11.2   Here I shall outline a very simple model of the optimum arrangement of roads and housing in the residential belt of a circular city. It gives an interesting economic application of the techniques of Chapters 9 and 10 in which the independent variable is not time.

The central business district of the city occupies a circle of given radius a. The residential belt spreads from there to a bigger circle of given radius b. We are also given the number of residents, N, and the amount of housing space to be allotted to each, $2\pi h$. The factor 2π is chosen to simplify notation later, and π is not a shadow price, but 3.14159 . . . , the ratio of the circumference of a circle to its diameter. So long as $2\pi hN < \pi(b^2 - a^2)$, there will be land left over for roads.

The roads are used by commuters to travel to the central business district and back, and the problem is to arrange roads and housing in such a way as to minimize the congestion costs of this travel. Suppose that we have a large number of evenly spaced radial roads, so that the trips necessary to reach the nearest one can be neglected, and attention concentrated on the trips radially to the central business district and back.

Suppose $N(r)$ residents live between r and b. Then there are $-N'(r)$ dr living in a small ring located between the radii r and $(r + dr)$, shown in Figure 11.2. These occupy $-2\pi hN'(r)$ dr units of land, leaving the rest out of the total area of $2\pi r$ dr in the ring to roads. Thus, along the circumference of the ring, such roads occupy a width of $2\pi[r + hN'(r)]$. Since these roads are used for commuting by the $N(r)$ residents living

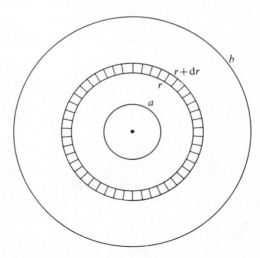

FIG. 11.2

farther away, the traffic density along them is proportional to $N(r)/[r + hN'(r)]$. There are $N(r)$ dr man-yards of travel at this density in each direction. Omitting factors of proportionality, suppose the congestion cost element contributed by these is

$$\{N(r)/[r + hN'(r)]\}^{k} N(r) \, \mathrm{d}r = N(r)^{k+1} \, \mathrm{d}r/[r + hN'(r)]^{k}$$

where k is a positive constant. In practice we find that k exceeds 2.

Thus our problem is to choose the function $N(r)$, subject to the end-point conditions $N(a) = N$ and $N(b) = 0$, to minimize

$$\int_a^b \frac{N(r)^{k+1}}{[r + hN'(r)]^k} \, \mathrm{d}r. \tag{11.10}$$

This is most easily tackled using the Euler-Lagrange equation. Simple differentiation gives

$$\frac{\mathrm{d}}{\mathrm{d}r} \left\{ \frac{-khN(r)^{k+1}}{[r + hN'(r)]^{k+1}} \right\} = \frac{(k+1)N(r)^k}{[r + hN'(r)]^k} \tag{11.11}$$

This looks quite formidable, but a simple substitution makes it manageable. Define $D(r) = N(r)/[r + hN'(r)]$, which is the traffic density at r except for a constant of proportionality. Then (11.1) becomes

$$-kh \, \mathrm{d}D^{k+1}/\mathrm{d}r = (k+1) \, D^k$$

or

$$\mathrm{d}D/\mathrm{d}r = -1/(kh). \tag{11.12}$$

This shows at once that traffic density in the optimum arrangement falls linearly with distance. In terms of an undetermined constant c_1, we have

$$D(r) = (c_1 - r)/(kh). \tag{11.13}$$

Recalling the definition of $D(r)$, we have a first-order differential equation for $N(r)$ which can be solved, introducing another constant of integration c_2. After some tedious calculation, it is possible to show that

$$N(r) = (c_1 - r)^{-k} \left\{ c_2 - \frac{1}{h} \int_a^r r(c_1 - r)^k \, \mathrm{d}r \right\} \tag{11.14}$$

The end-point conditions can then be used to determine c_1 and c_2. This can be computed numerically.

Interesting qualitative features can be found by examining the width of roads. Within a constant of proportionality, define $W(r) = r + hN'(r)$. Obviously we must have $W(r) \leqslant r$. We have the differential equation

$$W'(r) = [(k+1)W(r) - kr]/(c_1 - r). \qquad (11.15)$$

The solution can be examined geometrically, as in Figure 11.3. Clearly $W(b) = 0$, as there is no reason to provide roads for zero traffic. Now

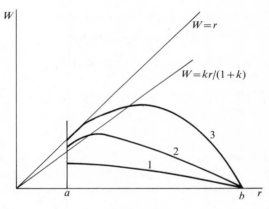

FIG. 11.3

$W'(r)$ has the same sign as $[W(r) - kr/(k+1)]$, and we can trace the solution backwards starting at b. We find three possibilities. The first has $W(r)$ decreasing all the way from a to b. This occurs when N is small relative to the maximum value it is allowed to attain, i.e. for sparsely populated cities. For more congested cities, we have the second case where $W(r)$ increases for a while and then decreases. For very heavily congested cities, the solution traced backward hits the line $W = r$, showing that there is a range near a where the entire area has to be devoted to roads. To solve this properly, of course, we have to formulate the problem allowing for such inequality constraints. This is more difficult, but the feature of the solution is clear. However, if a were a choice variable, we would not allow the third case to occur, for

it would be desirable to expand the central business district to occupy
the space that is being wasted on roads just outside a.

The last remark is just one example of the way in which the model
could be, and indeed has been, generalized. Once again, such
refinements must be left to be pursued by the interested readers. These
changes, however, leave the basic qualitative features of the allocation
of space to roads in the residential belt unchanged.

EXERCISES

11.1   Consider a firm which faces a demand curve

$$q(t) = a - x(t) - bp(t)$$

at time t. Here a and b are positive constants, $p(t)$ and $q(t)$ are
respectively the price and the quantity demanded at t, and $x(t)$ equals
the sales of its competitors at t. These sales are governed by the
differential equation

$$\dot{x}(t) = k[p(t) - p^*],$$

where k and p^* are positive constants. This shows that competitors
enter or expand if they see this firm charging a price above the 'limit
price' p^*. The average costs of production are constant and equal to c
at all times, and there is a constant rate of interest r. The firm wishes to
maximize

$$\int_0^\infty [p(t) - c]q(t)\, e^{-rt}\, dt,$$

with $x(0)$ given.

Apply the Maximum Principle to solve this problem, taking x as the
state variable and p as the control variable. Construct the diagram in
(x, p) space showing the possible solution paths of the appropriate
differential equations. Hence find the qualitative features of the
optimum pricing policy of the firm over time. Assume that $p^* \geqslant c$, and
obtain the conditions on the parameters of the problem which must be
satisfied if the competing firms retain positive sales in the limit.

11.2   An economy begins its planning at time 0, when it has a stock

S_0 of an exhaustible resource. It chooses a plan of depletion at a rate $R(t)$ as a function of time, subject to the feasibility requirement

$$\int_0^\infty R(t)\,\mathrm{d}t \leqslant S_0.$$

The plan aims to maximize

$$\int_0^\infty U(R(t))\,\mathrm{e}^{-\alpha t}\,\mathrm{d}t,$$

where

$$U(R) = R^{1-\epsilon}/(1-\epsilon),$$

α and ϵ being positive constants.

Taking $S(t)$, the stock of the resource that remains at t, as the state variable, and $R(t) = -\dot{S}(t)$ as the control variable, show that the multiplier $\pi(t)$ obtained from the Maximum Principle is constant over time. Hence deduce that the optimum depletion plan is given by

$$R(t) = (\alpha S_0/\epsilon)\,\mathrm{e}^{-(\alpha/\epsilon)t}.$$

Solve the problem using the Euler-Lagrange equation. Set up the Dynamic Programming equation and guess a solution to it as in Exercise 10.1.

What problems arise if α equals zero?

FURTHER READING

For relatively simple expositions of optimum saving theory, see
SOLOW, R. M. *Growth Theory: An Exposition,* Clarendon Press, Oxford, 1970, Chapter 5.
DIXIT, A. K. *The Theory of Equilibrium Growth,* Oxford University Press, 1976, chs. 5, 7.
For a more advanced treatment, see Intriligator, op. cit. (p. 68) ch. 16. Heal op. cit., chs. 12, 13. See also
WAN, H. Y. Jr. *Economic Growth,* Harcourt Brace Jovanovich, New York, 1971, chs. 9—11.
An important related branch is the theory of optimum depletion of exhaustible resources. See the special symposium issue of the *Review of Economic Studies,* 1974. Exercise 11.2 is the simplest case of the problem.

For richer models of the optimum size and organization of cities, see

MILLS, E. S. and DE FERRANTI, D. M. 'Market Choices and Optimum City Size', *American Economic Review,* LXI(2), May 1971, pp. 340–5.

DIXIT, A. K. 'The Optimum Factory Town', *The Bell Journal of Economics and Management Science,* 4(2), Autumn 1973, pp. 637–51.

For details of the problem of Exercise 11.1, see

GASKINS, D. W. Jr., 'Dynamic Limit Pricing: Optimal Pricing under Threat of Entry', *Journal of Economic Theory,* 3(3), September 1971, pp. 306–22, and comment by N. J. IRELAND, *Journal of Economic Theory,* 5(2), October 1972, pp. 303–5.

Models of optimum capital accumulation which use an objective more general than (11.2) can be found in

RYDER, H. E. Jr. and HEAL, G. M., 'Optimum growth with intertemporally dependent preferences', *Review of Economic Studies,* XL (1), January 1973, pp. 1–31.

WAN, H. Y. Jr., 'Optimal saving programs under intertemporally dependent preferences', *International Economic Review,* 11(2), October 1970, pp. 521–47.

Concluding Comments

I hope that I have provided sufficient theory and applications in this book to give the general economic theorist a good working understanding of optimization methods. However, readers who wish to specialize have large areas for further reading and thought available. The books by Malinvaud, Heal, Intriligator and Luenberger cited frequently in individual chapter reading lists will provide an excellent start. I have listed them here roughly in increasing order of mathematical sophistication.

One large and important area concerning optimization that I have omitted completely is that of decision-making under uncertainty. As with time, optimization under uncertainty does not introduce radically new basic theories, but the structure that arises when the various functions are expected values under some probability distributions leads to richer results. A useful treatment even at a very cursory level would add far too much to the bulk of this book. I shall therefore merely suggest some readings. These are

ARROW, K. J. *Essays in the Theory of Risk-bearing,* North-Holland, Amsterdam, 1970, especially chs. 1 and 3,

DE GROOT, M. H. *Optimal Statistical Decisions,* McGraw-Hill, New York, 1970,

and a forthcoming collection of articles edited by P. A. DIAMOND and M. ROTHSCHILD.

Index